Trieste Notes in Physics

P. Budinich A. Trautman

The Spinorial Chessboard

Springer-Verlag Berlin Heidelberg New York
London Paris Tokyo

Paolo Budinich

International School for Advanced Studies,
I-34014 Trieste, Italy

Andrzej Trautman

Instytut Fizyki Teoretycznej, Uniwersytet Warszawkski,
Hoza 69, PL-00681 Warszawa, Poland

ISBN-13: 978-3-540-19078-3 e-ISBN-13: 978-3-642-83407-3
DOI: 10.1007/978-3-642-83407-3

Library of Congress Cataloging-in-Publication Data. Budinich, P. (Paolo) The spinorial chessboard. (Trieste notes in physics) Bibliography: p. 1. Spinor analysis. I. Trautman, A. (Andrzej) II. Title. III. Series. QC20.7.S65B83 1988 530.1'5 88-12332

Softcover reprint of the hardcover 1st edition 1988

Offset printing and bookbinding: Druckhaus Beltz, D-6944 Hemsbach/Bergstr.
2156/3150-543210

Preface

The discovery of spin in 1925 and, in 1928, Dirac's derivation of the spinor field equation for the electron, prompted the theoretical prediction of the existence and properties of antimatter, only later confirmed experimentally. These successes, along with the basis of spinor calculus, developed by Pauli, Weyl and others, led many of the leading physicists of the time to conjecture that spinors might provide the clue for understanding the many obscure and apparently uncorrelated phenomena of quantum physics. In fact, in several later discoveries, including the exclusion principle, Fermi-Dirac statistics, weak interactions and parity violation, spinors did indeed play a major role, albeit not the revolutionary one expected for the conceptual foundations of physics by some great scientists such as Heisenberg.

The advent of relativity, gauge field theories and, more recently, of superfield and super-string theories have enhanced the role of Riemannian geometry in higher-dimensional spaces – and hence also that of spinors – in fundamental theoretical physics. Spinors may also be conceived from a geometrical point of view as elementary geometrical objects, as shown by Cartan, who discovered them and who stressed the equivalence of projective simple spinors of complex Euclidean spaces with null planes of maximal dimension in these spaces. The corresponding projective geometry, remarkably rich and elegant, was subsequently primarily the subject of mathematical studies. However, the recent renewed interest of physicists in the geometry of multidimensional spaces may lead to a revival of former ideas about the possible role of spinors in physics, in particular, exploiting the geometrical properties developed by Cartan and others.

This book represents a first step towards the study of spinors and simple spinors in higher-dimensional real pseudoeuclidean spaces that are of possible interest to physicists. Among useful techniques presented are Clifford algebras associated with real vector spaces with scalar products (in particular their representations, grading and periodicities), general properties of spinor spaces, bilinear forms and conjugations. Special attention is paid to research work in theoretical physics, where often, besides theorems and propositions, algorithms are required for explicit calculations.

This work was the result of collaboration between the Institute of Theoretical Physics of the University of Warsaw and the International School for Advanced Studies, Trieste. The authors wish to express their gratitude to those who made their stays in Warsaw and Trieste so pleasant.

We have had interesting discussions on the topics presented in these Notes with Mathias Blau, Michel Cahen, Ludwik Dabrowski, Paolo Furlan, Simone Gutt, Wojciech Kopczynski, André Lichnerowicz, Pertti Lounesto, Roger Penrose, Ryszard Raczka, Ivor Robinson, Walter Thirring, Engelbert Schucking and Helmut Urbantke. We thank Rosanna Sain and Sergio Stabile for their help in preparing the manuscript.

The research of one of us (A.T.) was supported in part by the National Science Foundation through Grant PHY 8513522 and the Polish Research Programme CPBP 01.03.

Trieste, Warszawa
January 1988

P. Budinich
A. Trautman

Contents

1. Introduction 1

1.1 A Little History 1
1.2 Null Elements and Simple Spinors 6
1.3 About the Present Work 10
1.4 Motivation and Outlook 13

2. Notation and Terminology 16

3. Vector Spaces and Inner Products 23

3.1 Complex Structure in a Real Vector Space 23
3.2 Quaternionic Structure in a Real Vector Space 24
3.3 Complex Conjugation and Hermitean Forms 26
3.4 Real and Quaternionic Structures in a Complex Vector Space . 29
3.5 Inner Products in Vector Spaces 31

4. Algebras and Their Representations 37

4.1 Definitions 37
4.2 Simple Algebras 41
4.3 Antiautomorphisms and Inner Products 42
4.4 Real Algebras 46
4.5 Graded Algebras 50

5. General Properties of Clifford Algebras 54

5.1 Definition and General Properties of Clifford Algebras 54
5.2 The Vector Space Structure of Clifford Algebras 57
5.3 The Graded Structure of Clifford Algebras 61
5.4 The Volume Element and Hodge Duality 66
5.5 Relation Between the Clifford Algebras of Vector Spaces of Adjacent Dimension 69

6. Complex Clifford Algebras 71

6.1 Dirac and Weyl Spinors 71
6.2 The Inner Products 73
6.3 Extensions of Representations 76
6.4 The Wall Groups 81

7. Real Clifford Algebras 84

7.1 The Index Periodicity 84
7.2 Charge Conjugation and Majorana Spinors 90
7.3 The Dirac Forms 100
7.4 Clifford Algebras of Euclidean Spaces 104
7.5 The Spinorial Chessboard 109
7.6 Summary 116

References 123

1. INTRODUCTION

Spinors — and structures associated with them — are among the geometrical notions whose importance was recognized as a result of research in physics. For a long time, the interest of physicists in spinors was restricted to three- and four-dimensional spaces (Euclidean and Minkowski). Spinors associated with them have two or four components. Recent work on fundamental interactions and their unification makes essential use of geometries of more than four dimensions. For this reason, spinor structures in higher dimensions and, in particular, Elie Cartan's "simple" or "pure" spinors, have now more chance of becoming relevant to physics than they had at the time of the appearance of the article by Brauer and Weyl (1935) and Cartan's (1938) lectures.

This set of notes contains a review of the first stage of our research oriented towards physical applications of spinors associated with higher-dimensional geometries. It is intended to be followed by an account of the spinor groups and structures, the geometry of simple spinors and twistors, and of the associated differential equations.

1.1 A little history

There is a prehistory of spinors: the period of time, before the discovery of the spin of the electron, when mathematicians considered notions and ideas closely related to those of spin representations (in the present day terminology). It begins probably with Leonhard Euler (1770) and Olinde Rodrigues (1840) who discovered new representations of rotations in three-dimensional space. The latter wrote an equation for a rotation $(x, y, z) \mapsto (x', y', z')$ equivalent to

$$X' = (1 + \tfrac{1}{4}(m^2 + n^2 + p^2)^{-1}\, U\, X\, U^\dagger \tag{1.1}$$

where

$$U = \begin{pmatrix} 1 + \frac{1}{2} ip & \frac{1}{2}(im + n) \\ \frac{1}{2}(im - n) & 1 - \frac{1}{2} ip \end{pmatrix}, \quad X = \begin{pmatrix} z & x - iy \\ x + iy & -z \end{pmatrix}, \tag{1.2}$$

and similarly for X'. The right hand side of (1.1) is rational in the components of the vector (m, n, p) parallel to the axis of rotation; the angle of rotation is $\omega = 2 \text{ arc tg } \frac{1}{2}\sqrt{m^2+n^2+p^2}$ and the unitary unimodular matrices $\pm U \cos \frac{1}{2}\omega$ cover the rotation in question. This may be interpreted to mean that Euler and Rodrigues knew that Spin(3) = SU(2). Formulae for rotations similar to (1.1) were also known to Carl Ludwig Gauss (cf. Cartan 1908).

The discovery of quaternions by William Rowan Hamilton (1844) led to a much simpler, "spinorial" representation of rotations: if $q = ix + jy + kz$ is a "pure" quaternion and u is a unit quaternion, then

$$q \rightarrow u\, q\, u^{-1}$$

is a rotation and every rotation can be so obtained. This observation, which can be used to establish the isomorphism Spin(3) = Sp(1), was made by Arthur Cayley (1845) who mentioned, however, that the result had been known to Hamilton. Cayley discovered also a quaternionic representation of rotations in four dimensions that was equivalent to the statement Spin(4) = Sp(1) x Sp(1) (Cayley 1855). Quaternions are now an important part of the structure of real Clifford algebras. In this context, it is instructive to recall the view of Lord Kelvin (quoted after Kline 1972):

> "Quaternions came from Hamilton after his really good work had been done; and though beautifully ingenious, have been an unmixed evil to those who have touched them in any way... Vector is a useless survival, or offshoot from quaternions, and has never been of the slightest use to any creature."

The Hamilton-Cayley representation of rotations in 3 and 4 dimensions by quaternions was generalized to higher-dimensional spaces by Rudolf O. Lipschitz (1886) who used for this purpose the associative algebras introduced by William K. Clifford (1878). The algebras considered by Clifford and Lipschitz were generated by n anticommuting "units" e_α with squares equal to -1. In E. Cartan's "Nombres complexes: Exposé, d'après l'article allemand de E. Study (Bonn)" there is a definition and classification of real Clifford algebras of arbitrary signature (Cartan 1908).

The road to spinors initiated by Euler and essentially completed by Clifford and Lipschitz may be described as being based on the idea of *taking the square root of a quadratic form*. Indeed the matrix X given by (1.2) is linear in x, y, z and has the property

$$X^2 = (x^2 + y^2 + z^2)\, I \qquad (1.3)$$

where I is the unit 2 by 2 matrix; Clifford algebras provide a universal method of generalizing (1.3) to higher dimensions and arbitrary signatures.

Spinors have another parentage, related to the study of representations of Lie groups and algebras. The Lie algebras of orthogonal groups have representations which do not lift ("integrate") to linear representations of the groups themselves. For example, the Lie algebra of

SO(3) is isomorphic to $\mathbb{R}^3$ with the vector product playing the role of the bracket,

$$[e_1, e_2] = e_3, \quad \text{etc.} \tag{1.4}$$

The representation of (1.4) given by $e_\alpha \to \sigma_\alpha/2i$, where the Pauli matrices are

$$\sigma_1 = \begin{pmatrix} 0 & 1 \\ 1 & 0 \end{pmatrix}, \quad \sigma_2 = \begin{pmatrix} 0 & -i \\ i & 0 \end{pmatrix}, \quad \sigma_3 = \begin{pmatrix} 1 & 0 \\ 0 & -1 \end{pmatrix}, \tag{1.5}$$

does not lift to a representation of SO(3), but integrates to a representation of SU(2), the simply-connected double cover of SO(3), or, in other words, to a two-valued representation of SO(3). Cartan (1913) determined all irreducible representations of the Lie algebras of the groups SO(n) and found that, for every $n > 2$, there are among them representations which do not lift to SO(n). This is so because the groups SO(n) are not simply-connected; the double valuedness comes from

$$\pi_1\,(SO(n)) = \mathbb{Z}_2 \quad \text{for} \quad n > 2 \tag{1.6}$$

and Spin(n) is the double cover of SO(n) which is simply-connected for $n > 2$. Cartan's approach was infinitesimal: he considered representations of Lie algebras only. Brauer and Weyl (1935) found global, spinorial representations of the groups Spin(n) for all n. This road to spinors may be called *topological:* it is related, in an essential way, to the non-triviality of the fundamental groups π_1 of the groups of rotations. It has the virtue of allowing a generalization of the notion of spinorial representations to general linear groups (Ne'eman 1978). As a manifold, the group $GL^+(n, \mathbb{R})$ of n by n real matrices with positive determinant is homeomorphic to the Cartesian product of manifolds,

$$SO(n) \times \mathbb{R}^{n(n+1)/2}. \tag{1.7}$$

Therefore, for $n > 2$, $\pi_1\,(GL^+(n, \mathbb{R})) = \mathbb{Z}_2$ and the group has a simply-connected universal cover $\widetilde{GL}^+(n, \mathbb{R})$ homeomorphic to

$$Spin(n) \times \mathbb{R}^{n(n+1)/2}. \tag{1.8}$$

The group $\widetilde{GL}^+(n, \mathbb{R})$, for $n > 2$, has no finite-dimensional faithful representations. In other words, spinors associated with the general linear group have an infinity of components. They have the virtue of not requiring, for their definition, any quadratic form or scalar product; they can be contemplated on a "bare" differentiable manifold without metric tensor. The topological approach to spinors is more general than the one based on the idea of linearization of a quadratic form.

The importance of the two-valued representations of the rotation group for physics became clear after the discovery of the intrinsic angular momentum — *spin* — of the electron (Uhlenbeck and Goudsmit 1925) and through the work of Wolfgang Pauli (1927), Paul A.M. Dirac (1928) and many other physicists on wave equations describing the behaviour of fermions, i.e. particles with half-integer spin. According to B.L. van der Waerden (1960), the name *spinor* is due to Paul Ehrenfest.

Hermann Weyl (1929) put forward a relativistic wave equation for massless particles described by a two-component spinor function. Weyl's equation was criticized by Pauli (1933) on the ground that it was not invariant under reflections. Ettore Majorana (1937) introduced another equation, closely related to Weyl's, based on a reality condition equivalent to the identification of the particle and its antiparticle. Two-component equations became accepted in elementary particle physics after the discovery of parity violation in weak interactions.

At first, spinors baffled physicists who, under the influence of relativity theory and despite Lord Kelvin's opinion, were becoming accustomed to scalars, vectors and tensors. In the words of C.G. Darwin (1928):

> "The relativity theory is based on nothing but the idea of invariance and develops from it the conception of tensors as a matter of necessity; and it is rather disconcerting to find that apparently something has slipped through the net, so that physical quantities exist, which it would be, to say the least, very artificial and inconvenient to express as tensors".

What is a spinor? Every physicist uses this notion frequently and knows it well, but amazingly diverse definitions of spinors are given in the literature. The differences among the definitions of spinors are more profound than those related to vectors and tensors; for spinors, there are differences in the substance and not only in the form of the definitions.

Geometry and physics require a scheme to deal with *fields* of quantities such as vectors, tensors and spinors. Tensors of various types are first defined in terms of vectors: for example, they may be described as multilinear maps on Cartesian products of vector spaces and their duals. This algebraic definition is then extended to differentiable manifolds by taking the tangent bundle and applying to it the "functor" corresponding to the type of tensors under study. No such functorial or natural construction can be given for spinors because there are topological obstructions to their existence on manifolds. Moreover, the "obvious" algebraic definition of a spinor space may be extended in inequivalent ways to manifolds (Trautman 1987). The algebraic definition, discussed in Ch. 5 to 7, may be formulated as follows (Chevalley 1954): assume, for simplicity, that V is a 2m-dimensional real vector space with a scalar product g_0. The space of (Dirac) spinors of (V, g_0) is the carrier space S_0 of a complex, faithful and irreducible representation of the Clifford algebra $C\ell(g_0)$. Since the algebra $C\ell(g_0)$ is simple, all such representations are equivalent and the 2^m-dimensional space S_0 is determined up to isomorphism.

There are at least two inequivalent extensions of the algebraic definition of spinors to manifolds. We recall them here for the special case of a 2m-dimensional oriented manifold M with a positive-definite Riemannian metric tensor g.

(i) The standard definition (Haefliger 1956, Borel and Hirzebruch 1958-60) of a spinor structure on M: it is a *spin prolongation* P of the bundle F_g of orthonormal frames of coherent orientation on M. There are bundle maps

$$\begin{array}{ccccc} & & \mathbb{Z}_2 & & \\ & & \downarrow & & \\ \mathrm{Spin}\,(2m) & \to & P & \to & M \\ \downarrow & & \downarrow & & \| \\ SO(2m) & \to & F_g & \to & M \end{array} \tag{1.9}$$

(see, for example, Dąbrowski and Trautman (1986) for details and references). The bundle $\Sigma \to M$ of Dirac spinors is associated with $P \to M$ by the standard representation of Spin (2m) in $S_0 = \mathbb{C}^{2^m}$. The prolongation P exists if, and only if, the second Stiefel-Whitney class of M vanishes.

(ii) If M admits an orthogonal almost complex structure J, then one can define a "Chevalley bundle"

$$S = \Lambda N \subset \Lambda(\mathbb{C} \otimes TM) \tag{1.10}$$

where N is the totally null subbundle of $\mathbb{C} \otimes TM$ consisting of all complex vectors of the form $u - i\,J(u)$, where $u \in TM$. The bundle $S \to M$ has S_0 as its typical fibre and there is a bundle map

$$C\ell(g) \times S \to S \tag{1.11}$$

making the fibre of $S \to M$ at $x \in M$ into the carrier space of a representation of the Clifford algebra $C\ell(g_x)$ associated with (T_xM, g_x), where g_x is the restriction of g to the tangent space T_xM.

The bundles Σ and S are inequivalent: among even-dimensional spheres only those of dimension 2 and 6 admit both Chevalley and Dirac bundles. The Dirac bundles of spheres are all trivial (Gutt 1986), but the Chevalley bundle of S_2 is not. All complex manifolds admit Chevalley bundles defined by their complex structure. In particular, this is true of the even-dimensional complex projective spaces which have no Dirac bundles.

For most purposes, one assumes the standard definition (i). We have mentioned definition (ii) to emphasize a certain non-uniqueness in the notion of spinors on manifolds. The latter definition is closely related to the approach to spinors through differential forms (Ivanenko and Landau 1928, Kähler 1960, Graf 1978) and to the representations of Clifford bundles considered by Karrer (1973).

1.2 Null elements and simple spinors

The approach to spinors exposed by Elie Cartan (1938) is based on the use of *null* [1)] (light-like, optical) geometrical elements: vectors with vanishing squares and linear spaces containing non-zero vectors orthogonal to the space. The connection between spinors and null elements is of fundamental importance for the applications of spinors in the theory of relativity (Penrose 1960, Penrose and Rindler 1984, 1986). It is at the basis of the Newman-Penrose (1962) formalism developed to study and solve Einstein's equations. The discovery of twistors by Penrose (1967) is closely linked to observations concerning a remarkable Robinson congruence of null lines in Minkowski space (Penrose 1987). Twistors have led to deep results, such as new methods for solving both linear and non-linear equations (Penrose and Mac Callum 1972, Ward 1977).

A connection between spinors and null vectors can be illustrated on the old problem of *Pythagorean triples,* i.e. triples x, y, z of positive integers such that

$$x^2 + y^2 = z^2 \tag{1.12}$$

Equation (1.12) means that the vector (x, y, z) is null with respect to a scalar product of signature (2.1). It is equivalent to the statement that the symmetric matrix

---

[1)] In pure mathematics the adjective "isotropic" is used to denote vectors with vanishing square and also vector spaces consisting of such vectors (Porteous 1981). Physicists refer to such objects as "null". The former choice is somewhat misleading since the word "isotropy" is often used in a different context: there is the isotropy subgroup defined by the action of a group in a space.

$$X = \frac{1}{2}\begin{pmatrix} z+y & x \\ x & z-y \end{pmatrix} \tag{1.13}$$

is of rank 1: det X = 0 and X ≠ 0. There thus exists a two-component real "spinor" (p, q) such that

$$X = \begin{pmatrix} p \\ q \end{pmatrix}(p, q) \tag{1.14}$$

or

$$x = 2pq, \quad y = p^2 - q^2, \quad z = p^2 + q^2 . \tag{1.15}$$

Not only does (1.15) give a solution of (1.12), but every Pythagorean triple of relatively prime integers (x, y, z) can be represented as in (1.15) by choosing a suitable couple of relatively prime integers p and q.

As an example closer to physics, consider the vectors **E** and **B** of a non-zero electromagnetic field, the complex vector

$$\mathbf{F} = \mathbf{E} + i\mathbf{B} = (F_1, F_2, F_3), \tag{1.16}$$

and the symmetric matrix

$$\Phi = \begin{pmatrix} F_1 + iF_2 & iF_3 \\ iF_3 & F_1 - iF_2 \end{pmatrix}. \tag{1.17}$$

Its determinant,

$$\det \Phi = F_1{}^2 + F_2{}^2 + F_3{}^2$$

vanishes if, and only if, the electromagnetic field is simple or null, i.e. when

$$\mathbf{E} \cdot \mathbf{B} = 0 \quad \text{and } \mathbf{E}^2 = \mathbf{B}^2. \tag{1.18}$$

If this is so, then there is a complex two-component spinor $\phi = \begin{pmatrix} \phi_1 \\ \phi_2 \end{pmatrix}$ such that

$$\Phi = \begin{pmatrix} \phi_1 \\ \phi_2 \end{pmatrix} (\phi_1, \phi_2).$$

The spinor $\phi \in \mathbb{C}^2$ is determined by F up to a sign and can be also used to form the Hermitean matrix

$$\psi = \begin{pmatrix} \phi_1 \\ \phi_2 \end{pmatrix} (\phi_1, \phi_2). \tag{1.19}$$

Equation (1.19) can be abbreviated to read $\psi = \phi\, \phi^\dagger$ and the matrix ψ represented as a linear combination of the three Pauli matrices and the unit matrix $\sigma_0 = I$,

$$\psi = k^\mu\ \sigma_\mu \quad \text{(summation over } \mu = 0, ..., 3) \tag{1.20}$$

The real vector $k \in \mathbb{R}^4$ with components given by (1.20) is null with respect to the Minkowski scalar product of signature (1,3). Moreover ,

$$k^o = |\,\mathbf{E}\,| = |\,\mathbf{B}\,| \quad \text{and} \quad k^o\, \mathbf{k} = \mathbf{E} \times \mathbf{B}\,,$$

where $(k^1, k^2, k^3) = -\mathbf{k}$. Simple electromagnetic fields characterized by (1.18) and (1.20) play a major role in the theory of shear free congruences of null geodesics in Lorentzian manifolds; they give rise to an "optical geometry" and a Cauchy-Riemann structure on the space of null geodesics (Robinson 1961, Penrose 1983a, Trautman 1985, Robinson and Trautman 1986).

To put in perspective these examples, consider the complex vector space $V = \mathbb{C}^{2m}$ with a scalar product g and a faithful irreducible representation

$$\gamma : C\ell(2m) \to \mathbb{C}(2^m) \tag{1.21}$$

of its Clifford algebra (these notions are defined and studied in detail in Chapter 6). Let $\phi \in S = \mathbb{C}^{2m}$ be a non-zero Dirac spinor. Its direction dir ϕ defines a vector subspace of V,

$$N(\text{dir}\ \phi) = \{u \in V \mid \gamma(u)\, \phi = 0\}. \tag{1.22}$$

From the basic property of the representation (1.21),

$$\gamma(u)\, \gamma(v) + \gamma(v)\, \gamma(u) = 2g\,(u, v), \tag{1.23}$$

it follows that N = N (dir ϕ) is *totally null*, i.e. every vector in N is null. The dimension of N is not larger than m. A necessary condition for N to be of the *maximal dimension* m is that ϕ be a *Weyl spinor*, i.e. an eigenvector of the helicity operator

$$\Gamma = i^m \gamma_1 \gamma_2 \cdots \gamma_{2m}, \tag{1.24}$$

where $\gamma_\alpha = \gamma(e_\alpha)$ and e_α (α = 1, ..., 2m) are the vectors of an orthonormal basis in V embedded in $C\ell(2m)$. This condition is also sufficient for m = 1, 2, and 3: there is a natural, bijective correspondence between the projective space of Weyl spinors and the set of maximal, totally null planes of the corresponding helicity (§ 5.4). For m ≥ 4 the complex dimension $2^{m-1}-1$ of the projective space of Weyl spinors is larger than the dimension m(m-1)/2 of the manifold

$$SO(2m)/U(m) \tag{1.25}$$

of maximal totally null planes. Elie Cartan calls a spinor *simple* (in the French edition, Cartan 1938; in the English translation, the adjective *pure* is used) if it defines by (1.22) a totally null plane of maximal dimension. Cartan shows that a Weyl spinor ϕ is simple if, and only if,

$$<B\,\phi, \gamma_{\alpha_1} \gamma_{\alpha_2} \cdots \gamma_{\alpha_p} \phi> = 0 \tag{1.26}$$

for all sequences of integers α such that

$$1 \le \alpha_1 < \alpha_2 < \ldots < \alpha_p \le 2m \quad \text{and} \quad 0 \le p \le m-1. \tag{1.27}$$

Here $B : S \to S^*$ is such that ${}^t\gamma_\alpha = B\,\gamma_\alpha\,B^{-1}$ and it is understood that for p = 0 condition (1.26) reduces to

$$<B\phi, \phi> = 0. \tag{1.28}$$

The m-form with components given by (1.26) for p = m characterizes the m-dimensional totally null plane associated with the simple spinor ϕ.

In eight dimensions (m = 4) equation (1.28) is the only condition for ϕ to be simple. Here simple spinors lie on a "null cone" in the eight-dimensional space of Weyl spinors; an interesting *triality*, or symmetry between the three eight-dimensional spaces (vector space and two spaces of Weyl spinors), appears in this case (Study 1903, Cartan 1925, Weiss 1933, Chevalley 1954, Tits 1959, Porteous 1981, Penrose and Rindler 1986).

Simple spinors can be defined in a similar manner for real vector spaces with a neutral scalar product. For other signatures, if one insists on staying within the domain of real numbers, the situation is much more complicated and subtle. For example, if the scalar product is positive-definite, then there are no null directions whatsoever and the group SO(n) of rotations acts transitively on the projective space $\mathbb{R}P_{n-1}$ of vector directions. For sufficiently high n, however, the action of Spin(n) on the projective spinor space is not transitive. The "simplicity" of a spinor can be measured by the dimension of its orbit under the action of the spin group: the lower the dimension, the simpler the spinor. Only partial results have been so far obtained on the classification of orbits of Spin (k, ℓ) and the geometrical interpretation of simple spinors in those cases (Porteous 1981, Igusa 1970, Popov 1977, Benn and Tucker 1988, Budinich 1986b, Budinich and Trautman 1986).

1.3 About the present work

In this paper, we describe in considerable detail the spinorial representations of the Clifford algebras associated with complex and real vector spaces. We give explicit methods to find the representations for arbitrary dimension and signature. We also present all the essential information about the invariant bilinear and Hermitean forms on the carrier spaces of the representations. Special attention is devoted to the appearance of Weyl and Majorana spinors (of two kinds), to charge conjugation and to the symmetry and signature of the invariant forms. Our main tool is the classical theorem about representations of simple algebras (§ 4.2).

To obtain an overall picture of the representations of Clifford algebras it is convenient to divide the study into several steps in such a way that at each step a new structure is introduced.

(i) At first, one forgets about the Clifford algebra everything but its structure of *algebra* $\mathcal{A}$. For any algebra $\mathcal{B}$, we denote by $2\mathcal{B}$ the direct sum $\mathcal{B} \oplus \mathcal{B}$, cf. § 4.5. There are two types of complex algebras,

$$\mathbb{C}(2^m) \quad \text{and } 2\mathbb{C}(2^m),$$

and five types of real algebras,

$$\mathbb{R}(2^m),\ 2\mathbb{R}(2^m),\ \mathrm{H}(2^m),\ 2\mathrm{H}(2^m) \text{ and } \mathbb{C}(2^m).$$

The integer m is simply related to the dimension of the underlying vector space. For example, considered as abstract algebras, the three algebras $\mathcal{C}\ell(4)$, $\mathcal{C}\ell(4,1)$ and $\mathcal{C}\ell(2.3)$ are all isomorphic to $\mathbb{C}(4)$.

(ii) If the Clifford algebra is considered together with its $\mathbb{Z}_2$-grading given by the main automorphism α, then there are still two types of complex algebras, but already eight classes of real algebras, cf. Table IV in § 7.1. This provides a classification finer than at the previous step, but one cannot determine the signature of the underlying vector space from the sole knowledge of its graded Clifford algebra $\mathcal{A}_0 \to \mathcal{A}$. For example, the graded algebra

$$2\mathbb{R}(8) \to \mathbb{R}(16)$$

is isomorphic to $C\ell_0(8,0) \to C\ell(8,0)$, $C\ell_0(4,4) \to C\ell(4,4)$ and $C\ell_0(0,8) \to C\ell(0,8)$. The class of the real algebra $C\ell(k, \ell)$ depends on

$$k - \ell \mod 8. \tag{1.28}$$

(iii) If

$$\gamma : \mathcal{A} \to \mathrm{End}\, S$$

is a faithful irreducible representation of a simple algebra $\mathcal{A}$ with an involutive antiautomorphism β, then the contragredient representation

$$\check{\gamma} : \mathcal{A} \to \mathrm{End}\, S^*, \quad \text{where } \check{\gamma}(a) = {}^t\gamma(\beta(a)),$$

is equivalent to γ and there exists an isomorphism $B : S \to S^*$ intertwining γ and $\check{\gamma}$. If $\mathcal{A}$ is central simple, then B is either symmetric or skew; it defines an inner product on S. The symmetry of B depends on the dimension n of the underlying vector space

$$ {}^tB = \begin{cases} B & \text{for } n = 0, 1, 2, 7 \mod 8 \\ -B & \text{for } n = 3, 4, 5, 6 \mod 8 \end{cases} \tag{1.29}$$

The double periodicity mod 8 given by (1.28) and (1.29) gives rise to a *chessboard* arrangement of real Clifford algebras alluded to in the title of this work and presented in detail in § 7.5 and Tables VI-IX.

(iv) There is a great wealth of structure in a Clifford algebra $\mathcal{A}$ taken together with the vector space V that generates it:

1. The natural linear isomorphisms

$$\mathcal{A} \simeq \Lambda\, V \simeq \Lambda\, V^* \tag{1.30}$$

allow an interpretation of elements of the Clifford algebra as multivectors or forms.

2. The grading, $\mathcal{A} = \mathcal{A}_0 \oplus \mathcal{A}_1$, may be used to define an associated graded or "super" Lie algebra. Its underlying vector space coincides with $\mathcal{A}$ and the graded bracket is

$$[a, b] = ab - (-1)^{pq}\, ba, \quad \text{where } a \in \mathcal{A}_p,\ b \in \mathcal{A}_q,$$

and p, q = 0 or 1. Of particular interest is the graded Lie subalgebra

$$\mathcal{L} = K \oplus V \oplus \Lambda^2 V. \tag{1.31}$$

If $u, v \in V$, then

$$[u, v] = uv + vu = 2g(u, v) \tag{1.32}$$

so that

$$[K, \mathcal{L}] = 0, \quad [V, V] \subset K, \quad [V, \Lambda^2 V] \subset V \tag{1.33}$$

and

$$[\Lambda^2 V, \Lambda^2 V] \subset \Lambda^2 V. \tag{1.34}$$

The last inclusion means that $\Lambda^2 V$ is an (ungraded) Lie subalgebra: it is the Lie algebra of the orthogonal and spin groups. These groups are also submanifolds of $\mathcal{A}$; we defer their detailed description to subsequent work.

3. If $\mathcal{B}$ is a minimal left ideal of a simple algebra with unity $\mathcal{A}$, then

$$\gamma : \mathcal{A} \to \text{End}\, \mathcal{B}, \quad \text{where } \gamma(a)\, b = ab,$$

for every $a \in \mathcal{A}$ and $b \in \mathcal{B}$, is a faithful irreducible representation of $\mathcal{A}$. This gives Chevalley's (1954) interpretation of spinors as elements of a minimal (left) ideal of a Clifford algebra.

All Clifford algebras are "supercentral" (§ 5.3). If (e_α) is an orthonormal basis for a scalar product of signature (k, ℓ), then the square of the volume element

$$\eta = e_1 e_2 \dots e_{k+\ell}$$

is

$$\eta^2 = (-1)^{(k-\ell)(k-\ell-1)/2}.$$

For $k-\ell \equiv 2$ or 3 mod 4 the square is negative and η belongs to the centre of $\mathcal{A}_0$ or $\mathcal{A}$, respectively. It may, therefore, be represented by i times the unit endomorphism of the space of Weyl or Dirac spinors.

There are at least two other "independent" ways of introducing complex numbers in quantum theory. The first comes from the observation that energy and momentum are related to translations. Infinitesimal translations are represented by first-order differential operators. To make them (formally) self-adjoint one has to multiply them by i. A related observation is that the Laplacian on compact Riemannian spaces ia a negative operator.

Another reason for considering complex wave functions and, in particular, spinor fields, has to do with electromagnetic interactions. According to the gauge, or "minimal interaction" principle, wave equations for charged particles contain the gradient operator d always in the combination d-ieA, where e is the charge and A the potential of the (external) electromagnetic field. The i comes from the fact that the Lie algebra of the group U(1) — the gauge group of electrodynamics — consists of pure imaginary numbers. It is not a trivial or obvious matter that the three i's (spinorial, quantum-mechanical and electromagnetic) are one and the same; but they are as indicated by the successes of the Dirac equation. Similar remarks have recently been made by Chen Ning Yang (1987).

1.4 Motivation and outlook

Every physicist will agree that spinors are a necessary and important tool in the description of fundamental interactions. The success of the Dirac equation is one of the most beautiful chapters of theoretical physics. Spinors play a major role in essentially all recent attempts at building new models (grand unification, supersymmetry, strings and membranes). They are also very useful in the classical, relativistic theory of gravitation (Penrose and Rindler, 1986). An impressive example of the usefulness of spinor analysis in a new domain has been provided by Edward Witten (1981) who proved the "positive energy theorem" in Einstein's theory in a manner which is more transparent than the earlier proof due to Schoen and Yau. Thirring (1972) showed that by spinors in a five-dimensional space one can obtain CP violation in a geometrical way. Recent renewal of interest in generalized Kaluza-Klein theories (cf., for example, the papers by Witten (1981), Abdus

Salam and J. Strathdee (1982), and Steven Weinberg (1983)) has led to considering spinors in spaces of dimension greater than four. In a somewhat different context, one of us (Budinich 1979, 1986b) proposed to consider fields of simple (pure) spinors in suitable higher-dimensional spaces and to relate them to wave-functions of physical particles. There are indications that in this manner a "natural" way of deriving interaction terms of Lagrangians of particles with internal symmetry may be obtained. Attempts have been made to write a differential equation for simple spinors, consistent with the quadratic constraints (1.26). For example, the method of Lagrange multipliers, applied to a variational principle in 7 space-time dimensions, leads to a Weyl equation for simple spinors with a "mass term" induced by the constraint (1.28) (cf. Budinich and Trautman 1986 and the references given there). Remarks on the possible physical relevance of simple spinors have also been made by Benn and Tucker (1984) and by A.D. Helfer (1983).

There are some "unexpected" applications of spinors: spinor connections on low-dimensional spheres coincide with simple, topologically non-trivial gauge configurations (Budinich and Trautman 1985). Spinors provide a fine tool for the study of topological properties of manifolds (Atiyah, Bott and Shapiro 1964, Atiyah and Singer 1968). There is a remarkable "spinorial" form of the Enneper-Weierstrass formula for solutions of the equation for minimal surfaces and of its extension to strings (Budinich 1986, Budinich and Rigoli 1987, and the references given there). It is based on a representation of complex and real null vectors in terms of spinors, analogous to those described in § 1.2.

Considerations such as these convince us that there may be something more to spinors than has been said and seen so far. This view has been put forward, quite a long time ago, by Roger Penrose who pursued the most comprehensive and farthest reaching programme of applying spinors — and their close relatives, twistors — in fundamental physics. We share his view "that we have still not yet seen the full significance of spinors — particularly the 2-component ones — in the basis structure of physical laws" (Penrose 1983b). We are inclined, however, to extend the belief in the significance of spinors to those associated with higher-dimensional geometries and replace the phrase about the 2-component spinors by one referring to simple spinors and the homogeneous spaces mentioned in § 1.2. (Note that, in four-dimensions, simple spinors have two components. More generally, Weyl spinors are simple in neutral spaces of dimension ≤ 6. In particular, twistors are simple).

Our work is an attempt to follow this road. The present article is a preparation for a systematic study of the spin and pin groups and of their representations in relation to simple spinors. We intend to make more precise the idea that the dimension of the orbit is a measure of the simplicity of spinors it contains, use our Proposition 4.2 to derive the biquadratic spinor identities (Case 1955), study (simple) spinor fields on homogeneous spaces — such as the ones arising from conformal compactification — and consider the possibilities offered by various schemes of dimensional reduction. As many before us, we draw encouragement from the Great Masters. Some of them have already been mentioned. We conclude this introduction with a quotation from Hermann Weyl (1946):

> "The orthogonal transformations are the automorphisms of Euclidean vector space. Only with the spinors do we strike that level in the theory of its representations on which Euclid himself, flourishing ruler and compass, so deftly moves in the realm of geometric figures".

2. NOTATION AND TERMINOLOGY

In this work we essentially adhere to the standard notation prevalent in mathematical physics. In a few cases, where the customs of mathematicians and physicists diverge, we have had to make painful decisions to achieve a consistent and clear notation. Most of our conventions are summarized below; some of them are explained as they appear in subsequent chapters.

We use the customary set-theoretical notation: the symbols $\cap$, $\cup$ and $\subset$ denote the intersection, union and inclusion of sets, respectively; $\emptyset$ is the empty set and $x \in X$ reads: x belongs to X. If X is a set, then

$$\{x \in X \mid P(x)\}$$

is the subset of X consisting of all those elements which have the property P. The Cartesian product $X \times Y$ of the sets X and Y consists of all pairs (x, y) such that $x \in X$ and $y \in Y$. A finite set may be described by enumerating its elements; if there are n of them, then the set may be represented as $\{x_1, x_2, ..., x_n\}$. A subset R of $X \times X$ is an equivalence relation in X if, for every $x, y, z \in X$ the following are true

$$(x, x) \in R; \text{ if } (x, y) \in R, \text{ then } (y, x) \in R;$$
$$\text{if } (x, y) \in R \text{ and } (y, z) \in R, \text{ then } (x, z) \in R.$$

If R is an equivalence relation in X, then the quotient set of X by R is

$$X/R = \{Y \subset X \mid Y \times Y \subset R \text{ and } (x \in Y \text{ and } (x, y) \in R) \Rightarrow y \in Y\}.$$

The canonical map

$$\kappa : X \rightarrow X/R$$

assigns to $x \in X$ the set $Y \in X/R$ containing x.

The letters $\mathbb{Z}$, $\mathbb{R}$, $\mathbb{C}$ and H denote the sets of integers, real numbers, complex numbers and quaternions, respectively. Addition and multiplication of numbers define in the set $\mathbb{Z}$ the structure of a ring. If m and $n \in \mathbb{Z}$ and m-n is divisible by $p \in \mathbb{Z}$, $p > 0$, then we say that m and n are congruent modulo p and write

$$m \equiv n \mod p.$$

The relation R_p in $\mathbb{Z}$, where

$$R_p = \{(m, n) \in \mathbb{Z} \times \mathbb{Z} \mid m \equiv n \mod p\},$$

is an equivalence relation and the quotient $\mathbb{Z}_p = \mathbb{Z} / R_p$ is also a ring; if p is a prime - and only in this case - it is a field, i.e. a commutative ring with all non-zero elements invertible. The sets $\mathbb{R}$ and $\mathbb{C}$ are also fields and H is a "non-commutative field": it is an associative, but non-Abelian, ring with all non-zero elements invertible.

If X and Y are sets, then

$$f : X \to Y$$

means "f is a map (function) from X to Y" and

$$X \ni x \mapsto f(x) \in Y$$

means "$f(x) \in Y$ is the value of f at $x \in X$". The composition of maps $f : X \to Y$ and $g : Y \to Z$ is the map $g \circ f : X \to Z$. For every set X, there is the identity map $id_X : X \to X$ such that $id_X(x) = x$ for every $x \in X$. One often writes id instead of id_X.

Most of the time, we shall work in this article with finite-dimensional vector spaces over the field K of real or complex numbers. If V is a vector space, then the identity map id_V is usually denoted by I. If V and W are vector spaces over K, then the map

$$f : V \to W$$

is said to be K-linear, or a homomorphism of vector spaces, if

$$f(\lambda u + \mu v) = \lambda f(u) + \mu f(v)$$

for every $u, v \in V$ and $\lambda, \mu \in K$. The set

$$\mathrm{Hom}_K(V, W) = \{f : V \to W \mid f \text{ is } K\text{-linear}) \tag{2.1}$$

is also a vector space over K. Homomorphisms of V into V are called endomorphisms of V, and

$$\mathrm{End}_K V = \mathrm{Hom}_K(V, V).$$

The symbol of the field is omitted whenever this does not lead to misunderstandings. If $f \in \mathrm{Hom}(V, W)$ and $v \in V$, then one often writes

$$\langle f, v\rangle \quad \text{instead of } f(v) \tag{2.2}$$

This convenient notation is similar to, but not identical with, the one introduced by Dirac (1958) for Hilbert spaces: the evaluation map

$$\mathrm{Hom}_K(V, W) \times V \to W$$

given by

$$(f, v) \mapsto \langle f, v\rangle$$

is bilinear. The dual of V is the vector space

$$V^* = \mathrm{Hom}_K(V, K); \tag{2.3}$$

its elements are called linear forms or 1-forms on V. If $f : V \to W$ is a homomorphism, then its transpose

$${}^t f : W^* \to V^*$$

is defined by

$$\langle {}^t f(\alpha), v\rangle = \langle \alpha, f(v)\rangle \tag{2.4}$$

for every $\alpha \in W^*$ and $v \in V$. The symbol $\circ$ is often omitted when composition of homomorphisms is considered. If $g : U \to V$ is another homomorphism, then ${}^t g : V^* \to U^*$ and

$${}^t(fg) = {}^t g\, {}^t f\,. \tag{2.5}$$

The tensor product of two finite-dimensional vector spaces V and W over K may be defined by

$$V \otimes_K W = \mathrm{Hom}_K(V^*, W). \tag{2.6}$$

If $v \in V$ and $w \in W$, then their tensor product $v \otimes w$ is an element of $\mathrm{Hom}(V^*, W)$ such that

$$\langle v \otimes w, \alpha\rangle = \langle \alpha, v\rangle\, w \tag{2.7}$$

for every $\alpha \in V^*$. The spaces $U \otimes (V \otimes W)$ and $(U \otimes V) \otimes W$ are isomorphic in a natural way; therefore $U \otimes V \otimes W$ is meaningful.

If (e_α), $\alpha = 1, \ldots, n$, is a linear basis in an n-dimensional vector space V, then every vector $u \in V$ can be written as

$$u = u^\alpha e_\alpha$$

The Einstein summation convention over repeated indices is assumed to hold in the last formula. With any linear basis (e_α) in V there is associated the dual basis (e^α) in V^* such that

$$\langle e^\alpha, e_\beta \rangle = \delta^\alpha{}_\beta \qquad (\alpha, \beta = 1, \ldots, n) \tag{2.8}$$

where

$$\delta^\alpha{}_\beta = 1 \text{ for } \alpha = \beta \text{ and } \delta^\alpha{}_\beta = 0 \text{ otherwise.}$$

The components of u with respect to (e_α) are given by $u^\alpha = \langle e^\alpha, u \rangle$. If (e_α), $\alpha = 1, \ldots, n$, and (f_μ), $\mu = 1, \ldots, m$, are linear bases in the vector spaces V and W, respectively, then the set of mn elements of the form

$$e_\alpha \otimes f_\mu$$

is a linear basis in $V \otimes W$. Let $F \in \text{End } V$, then

$$F(e_\alpha) = F_\alpha{}^\beta e_\beta$$

and then n^2 numbers $F_\alpha{}^\beta$ $(\alpha, \beta = 1, \ldots, n)$ are the components of F with respect to (e_α). The trace

$$\text{Tr } F = F_\alpha{}^\alpha \tag{2.9}$$

does not depend on the choice of the basis and

$$\text{Tr } FG = \text{Tr } GF \tag{2.10}$$

for every $F, G \in \text{End } V$.

If $f \in \text{Hom}(V, W)$ is invertible, then f is said to be an isomorphism. An automorphism of V is an invertible endomorphism of V. The set of all automorphisms of V forms the general linear group GL(V) of the vector space V. In particular,

$$GL(n, K) = GL(K^n) \quad \text{for } K = \mathbb{R} \text{ or } \mathbb{C}.$$

The direct sum of vector spaces V and W is the vector space $V \oplus W$ consisting of all pairs (v, w), where $v \in V$ and $w \in W$. The operations are defined by $(v, w) + (v', w') = (v+v', w+w')$ and $\lambda(v, w) = (\lambda v, \lambda w)$ for every $\lambda \in K$; $v, v' \in V$ and $w, w' \in W$. There are natural identifications such as

$$(V_1 \oplus V_2) \otimes W = (V_1 \otimes W) \oplus (V_2 \otimes W),$$

$$(V \oplus W)^* = V^* \oplus W^*, \text{ etc.}$$

The notation of a direct sum can be extended to the case of several summands.

If V is a vector subspace of W, then there is the equivalence relation R in W

$$R = \{(u, v) \in W \times W \mid u-v \in V\}$$

and the quotient set W/R can be given the structure of a vector space in a natural manner; this quotient vector space is denoted by W/V. If $f \in \mathrm{Hom}(W, U)$ and $f(v) = 0$ for every $v \in V$, then f passes to the quotient: there is homomorphism $F : W/V \to U$ such that $f = F \circ \kappa$ where $\kappa : W \to W/V$ is the canonical map.

There is a point where our notation and terminology differ from those accepted by the majority of physicists: it concerns complex conjugation and "antilinear" maps. If μ is a complex number, then $\bar{\mu}$ is the conjugate number. If V and W are complex vector spaces, i.e. vector spaces over $\mathbb{C}$, then a map

$$f : V \to W$$

is said to be semi-linear (by physicists: "antilinear") if

$$f(u + v) = f(u) + f(v) \quad \text{and} \quad f(\mu u) = \bar{\mu}\, f(u)$$

for every $\mu \in \mathbb{C}$ and $u, v \in V$. If U, V, W are complex vector spaces, then a map

$$f : U \times V \to W$$

is said to be sesquilinear if

$$u \to f(u, v)$$

is semi-linear in u for every $v \in V$ and

$$v \to f(u, v)$$

is linear in v for every $u \in U$. For example, the Hermitean scalar product in a complex Hilbert space V is a sesquilinear map $V \times V \to \mathbb{C}$.

The tensor algebra of a vector space V over K

$$\mathcal{T}(V) = \bigoplus_{k=0}^{\infty} \overset{k}{\otimes} V \tag{2.11}$$

is defined by putting

$$\overset{o}{\otimes} V = K, \quad \overset{k}{\otimes} V = V \otimes \ldots \otimes V \quad \text{(k factors)}$$

and giving an associative multiplication which can be described as follows. An element t of $\mathcal{T}(V)$ is a sequence $t = (t_0, t_1, \ldots)$ such that

$$t_k \in \overset{k}{\otimes} V \quad \text{and} \quad t_k = 0 \text{ for almost all k.}$$

If s is another such sequence, then their tensor product is $t \otimes s$, where

$$(t \otimes s)_k = \sum_{p+q=k} t_p \otimes s_q$$

is a finite sum. Let $\mathcal{S}(V)$ be the two-sided ideal in $\mathcal{T}(V)$ generated by all elements of the form $v \otimes v$, where $v \in V$. The quotient algebra

$$\Lambda V = \mathcal{T}(V) / \mathcal{S}(V) \tag{2.12}$$

is called the Grassmann algebra of V. Let

$$\kappa : \mathcal{T}(V) \to \Lambda V$$

be the canonical map; the multiplication (exterior or wedge product) in Λ V is defined by

$$\kappa(t) \wedge \kappa(s) = \kappa(t \otimes s)$$

for every t, s $\in \mathcal{T}(V)$. Since κ is injective when restricted to $K \oplus V$, one can identify this space with its image in ΛV so that

$$v \wedge v = 0 \quad \text{for every } v \in V.$$

Moreover,

$$\Lambda V = \bigoplus_{k=0}^{n} \Lambda^k V \tag{2.13}$$

where

$$\Lambda^k V = \kappa \left(\overset{k}{\otimes} V \right)$$

and it is clear that the latter space reduces to the zero vector for $k > n = \dim_K V$. If (e_α) is a basis of V, then the set of $\binom{n}{k}$ sequences

$$e_{\alpha_1} \wedge e_{\alpha_2} \wedge \ldots \wedge e_{\alpha_k} , \tag{2.14}$$

where $1 \leq \alpha_1 < \alpha_2 < \ldots < \alpha_k \leq n$, is a basis of $\Lambda^k V$. Therefore, the Grassmann algebra of an n-dimensional vector space is 2^n-dimensional.

3. VECTOR SPACES AND INNER PRODUCTS

Clifford algebras and spinor spaces occurring in physics are vector spaces over the real or complex numbers. Quaternions also appear, in a natural manner, as was alluded to in the Introduction. There are subtle relations between these number fields and the signature of the quadratic form under consideration. In some cases, there is a "charge conjugation" which allows the definition of real spinors. To prepare ground for a systematic presentation of such matters, we summarize here some elementary notions related to the introduction of real, complex and quaternionic structures in vector spaces. We also review the definitions and basic properties of inner products and Hermitean forms needed in the sequel.

3.1 Complex structure in a real vector space

Let W be a 2n-dimensional real vector space. A linear map $J : W \to W$ such that $J^2 = -\mathrm{id}$ is said to define a *complex structure* in W. Given such a J one can make W into an n-dimensional complex vector space by defining

$$(a + \sqrt{-1}\, b)v = av + b\, J(v)$$

for any $a, b \in \mathbb{R}$ and $v \in W$.

For example, if V is an n-dimensional real vector space, then the direct sum $W = V \oplus V$ can be given a complex structure by putting

$$J\,(u,v) = (-v,u) \quad \text{for any } u, v \in V.$$

The n-dimensional complex vector space W is then said to be the *complexification* of V; instead of (u, v) and $V \oplus V$ one often writes $u + \sqrt{-1}\, v$ and $V + \sqrt{-1}\, V$, respectively. Alternatively — and equivalently — one can view the complexification of V as consisting in taking the tensor product $\mathbb{C} \otimes_{\mathbb{R}} V$.

Let W_1 and W_2 be two real, even-dimensional vector spaces with complex structures J_1 and J_2, respectively. An $\mathbb{R}$-linear map

$$f : W_1 \to W_2$$

is also $\mathbb{C}$-linear if, and only if,

$$f\, J_1 = J_2\, f.$$

In particular, let $W_1 = W_2 = W$ be real 2n-dimensional with complex structure J. If $(e_1,...,e_n)$ is a basis in W considered as a complex space, then

$$(e_1, ..., e_n, J(e_1), ..., J(e_n))$$

is a basis in W considered as a real space. With respect to this basis, the endomorphism J is represented by the matrix

$$\begin{pmatrix} 0 & -I \\ I & 0 \end{pmatrix}$$

where I denotes the n x n unit matrix. An $\mathbb{R}$-linear endomorphism f of W is also $\mathbb{C}$-linear iff it commutes with J, i.e. iff its components are represented by a matrix of the form

$$\begin{pmatrix} a & -b \\ b & a \end{pmatrix}$$

where $a, b \in \mathbb{R}(n)$. With respect to the complex basis $(e_1, ..., e_n)$, the same endomorphism is represented by the matrix

$$a + \sqrt{-1}\, b \in \mathbb{C}(n).$$

Here $\mathbb{R}(n)$ and $\mathbb{C}(n)$ denote the sets of all n by n matrices with real and complex entries, respectively.

3.2 Quaternionic structure in a real vector space

Consider a 4n-dimensional real vector space V and a couple (J,K) of endomorphisms of V such that

$$JK + KJ = 0 \quad \text{and} \quad J^2 = K^2 = -\mathrm{id}. \tag{3.1}$$

Such a couple defines a *quaternionic structure* in V by making this set into a *right H-module* in the following sense: for every quaternion

$$q = t + ix + jy + kz \qquad (t, x, y, z, \in \mathbb{R})$$

and vector $v \in V$ there is defined their product vq by

$$vq = t + xJK(v) + yJ(v) + zK(v) \tag{3.2}$$

so that

$$v(q_1 + q_2) = vq_1 + vq_2, \quad (v+w)q = vq + wq$$

and

$$v(tq) = (tv)q, \quad (vq_1)q_2 = v(q_1 q_2)$$

for every $v, w \in V$; $q_1, q_2 \in H$ and $t \in \mathbb{R}$.

A *quaternionic endomorphism* of V is an $\mathbb{R}$-linear endomorphism f of V commuting with both J and K. The latter condition is equivalent to

$$f(vq) = f(v)q \text{ for every } v \in V \text{ and } q \in H. \tag{3.3}$$

Note that the endomorphisms J and K are *not* quaternionic.

Let $(e_1, \ldots, e_n)$ be a basis in V considered as a right H-module; the components v^μ ($\mu = 1, \ldots, n$) of a vector

$$v = e_\mu v^\mu$$

are quaternions. Let f and g be quaternionic endomorphisms of V and put

$$f(e_\mu) = e_\nu f^\nu{}_\mu \text{ and } g(e_\mu) = e_\nu g^\nu{}_\mu.$$

Then the composition g f is also a quaternionic endomorphism and

$$(gf)(e_\mu) = e_\nu g^\nu{}_\rho f^\rho{}_\mu$$

The assignment

$$f \mapsto (f^\nu{}_\mu)$$

is an isomorphism of the algebra of quaternionic endomorphisms of V on the matrix algebra H(n).

3.3 Complex conjugation and Hermitean forms

Recall that a complex vector space S consists of a set (of vectors), usually denoted by the same letter S, and two maps (operations): addition of vectors and multiplication of vectors by complex numbers. Let us denote — for a short while — the result of multiplication of $\phi \in S$ by $\lambda \in \mathbb{C}$ as

$$\text{prod}\,(\phi, \lambda, S)$$

to be read as "product of ϕ by λ in S". The two operations satisfy a number of well-known axioms. We now define the *complex conjugate* vector space $\bar{S}$ to consist of the same set of vectors as that of S, to have the same addition of vectors as in S and to be given the conjugate multiplication law:

$$\text{prod}\,(\phi, \lambda, \bar{S}) = \text{prod}\,(\phi, \bar{\lambda}, S) \tag{3.4}$$

The axioms of a complex vector space are easily seen to be saitsfied in $\bar{S}$. Let us now agree to *denote by* $\bar{\phi}$ *the vector* ϕ *when it is considered as an element of the complex conjugate space* $\bar{S}$. Then (3.4) is equivalent to

$$\text{prod}\,(\bar{\phi}, \bar{\lambda}, \bar{S}) = \overline{\text{prod}\,(\phi, \lambda, S)} \tag{3.5}$$

We can now revert to the traditional notation for products of vectors by numbers and write

$$\bar{\lambda}.\bar{\phi} = \overline{\lambda\,\phi} \tag{3.6}$$

instead of (3.5); there can be no confusion because the bar over ϕ on the left-hand-side forces us to consider the vector as belonging to $\bar{S}$ and use the corresponding multiplication law. On the right-hand-side, on the contrary, one first multiplies ϕ by λ in S and then puts the resulting vector in the barred basket. Eq.(3.6) means that the *bar map* $\phi \mapsto \bar{\phi}$ is *semi-linear*: it is in fact a semi-linear isomorphism of S on $\bar{S}$. It is clear that $\bar{\bar{S}}$ may be identified with S and $\bar{\bar{\phi}}$ with ϕ.

If $f : S \to T$ is a linear map of complex vector spaces, then the linear map

$$\bar{f} : \bar{S} \to \bar{T}$$

is defined by

$$\bar{f}(\bar{\phi}) = \overline{f(\phi)} \tag{3.7}$$

If $g : T \to U$ is another such map, then

$$\overline{g \circ f} = \bar{g} \circ \bar{f}. \tag{3.8}$$

Note that if f is linear, then $\bar{f}$ is also linear, but the correspondence $f \mapsto \bar{f}$ is semi-linear, $\overline{\lambda f} = \bar{\lambda}\bar{f}$, $\lambda \in \mathbb{C}$. The bar map is a universal semi-linear map in the sense that if $f : S \to T$ is semi-linear, then $\phi \mapsto \overline{f(\phi)}$ and $\bar{\phi} \mapsto f(\phi)$ are linear maps: every semi-linear map is the composition of a linear one with the bar map.

For every complex vector space S, the spaces

$$(\bar{S})^* = \{\alpha : \bar{S} \to \mathbb{C} \mid \alpha \text{ is linear}\} \tag{3.9}$$

and

$$\overline{(S^*)} = \{\bar{\beta} \mid \beta : S \to \mathbb{C} \text{ is linear}\} \tag{3.10}$$

are, in a natural manner, isomorphic to each other and to the space of semi-linear forms on S. The isomorphism $\iota : (\bar{S})^* \to \overline{(S^*)}$ is given by

$$\iota(\alpha) = \bar{\beta}, \quad \text{where } <\beta, \phi> = \overline{<\alpha, \bar{\phi}>} \quad \text{and } \phi \in S.$$

If $\alpha : \bar{S} \to \mathbb{C}$ is linear, then $\phi \mapsto <\alpha, \bar{\phi}>$ is semi-linear. The existence of the isomorphism ι justifies identifying the space $(\bar{S})^*$ with $\overline{(S^*)}$ and denoting it $\bar{S}^*$.

If $f : S \to T$ is a linear map of complex vector spaces, then similar considerations allow the identification of ${}^t(\bar{f})$ with $\overline{({}^t f)}$. This *Hermitean conjugate map*

$${}^t\bar{f} : \bar{T}^* \to \bar{S}^*$$

is sometimes denoted by $f^\dagger$. If $g : T \to U$ is another linear map, then

$$(gf)^\dagger = f^\dagger g^\dagger. \tag{3.11}$$

In particular, a linear map

$$f : \bar{S} \to S^*$$

is said to be *Hermitean* if

$$f^\dagger = f\,. \tag{3.12}$$

If f is Hermitean, then the sesquilinear form

$$f : S \times S \to \mathbb{C}$$

defined by

$$f(\phi, \psi) = \langle f(\bar{\phi}), \psi \rangle \tag{3.13}$$

is Hermitean in the classical sense,

$$f(\psi, \phi) = \overline{f(\phi, \psi)} \tag{3.14}$$

for every $\phi, \psi \in S$.

Given a basis (e_μ), $\mu = 1, \ldots, n$, in the n-dimensional complex vector space S, one can express any vector $\phi \in S$ as $\phi = \phi^\mu e_\mu$, where the complex numbers ϕ^μ are the components of ϕ with respect to (e_μ). The correspondence

$$\phi \mapsto \phi^\mu = \langle e^\mu, \phi \rangle$$

is linear and defines a basis (e^μ) in S*, called the *dual basis*. Similarly, by virtue of (3.6)

$$\bar{\phi} = \bar{\phi}^\mu \, \bar{e}_\mu$$

where $(\bar{e}_\mu)$ is the basis in $\bar{S}$ consisting of the same vectors as (e_μ). The linear correspondence

$$\bar{\phi} \mapsto \bar{\phi}^\mu$$

defines a basis $(\overline{e^\mu})$ in $\bar{S}^*$;

$$\overline{\phi^\mu} = \langle \overline{e^\mu}, \bar{\phi} \rangle \; .$$

If f is an endomorphism of S, then

$$f(e_\mu) = f^\nu{}_\mu \, e_\nu$$

where $\mu, \nu = 1, \ldots, n$ and the components $f^\nu{}_\mu$ of f with respect to (e_μ) are complex numbers. For the barred endomorphism $\bar{f}$ of $\bar{S}$ we have, on the basis of (3.6) and (3.7),

$$\bar{f}(\bar{e}_\mu) = \overline{f^\nu}_\mu \, \bar{e}_\nu .$$

In the spinor calculus originated by B.L. van der Waerden (1929) it is customary to replace the bars over the basis vectors and the components of geometric objects by dots put above the indices. For typographical reasons, R. Penrose (1960) replaces the dots by primes. Thus a generic element of $\bar{S}$ is represented as $\phi^{\dot{\mu}} e_{\dot{\mu}}$. This notation is particularly convenient when one considers — as one often does — spinors that are elements of tensor products of the four spaces S, S*, $\bar{S}$ and $\bar{S}^*$. For example, the components of an element of $S \otimes \bar{S}$ may be denoted by $\psi^{\mu\dot{\nu}}$.

3.4 Real and quaternionic structures in a complex vector space

In § 3.1 we recalled how a complex structure can be introduced in a real vector space. Let us now consider the problem of building a real space from a complex one; this question is relevant to the construction of Majorana spinors.

Let us first note that every complex vector space S has a *real form* obtained by taking the same vectors as in the original space and restricting the scalars to be real. If $\phi \in S$ and $\phi \neq 0$ then the vectors ϕ and $\sqrt{-1}\,\phi$ are linearly independent in the real form of S; therefore

$$\dim_{\mathbb{R}} S = 2 \dim_{\mathbb{C}} S$$

Another problem is to represent S as the direct sum of two real spaces, the "real and imaginary parts of S". There is no canonical way of doing this: such a splitting is an additional *real structure* in S. It may be introduced as follows. Let

$$C : S \to \bar{S} \tag{3.15}$$

be a linear map such that

$$\bar{C}C = \mathrm{id} \tag{3.16}$$

Any vector $\phi \in S$ can be written as

$$\phi = \phi^+ + \phi^-$$

where

$$\phi^{\pm} = \frac{1}{2}(\phi \pm \bar{C}\bar{\phi})$$

Therefore, there is a direct sum decomposition,

$$S = S^+ \oplus S^- \qquad (3.17)$$

where

$$S^{\pm} = \{\phi \in S \mid \bar{\phi} = \pm\, C\phi\}.$$

Both S^+ and S^- are real vector spaces and $K : S^+ \to S^-$, where $K(\phi) = \sqrt{-1}\,\phi$, is an isomorphism. Therefore,

$$\dim_{\mathbb{R}} S^+ = \dim_{\mathbb{R}} S^- = \dim_{\mathbb{C}} S.$$

Given an "abstract" complex vector space S, there is no "natural" decomposition of S into a direct sum (3.17). This has interesting "global" consequences. Consider, as an example, the tangent bundle of a 2-dimensional, oriented sphere. Each of the tangent spaces is real 2-dimensional and has a "natural" complex structure, defined by J = rotation of vectors by 90° in agreement with the preferred orientation. The tangent bundle is thus a complex line bundle, but it does not admit a smooth real structure. Such a structure would be equivalent to giving a smooth field of directions in the tangent spaces to the sphere.

Consider now a complex vector space S with a linear map (3.15) such that

$$\bar{C}C = -\mathrm{id} \qquad (3.18)$$

By taking the determinant of both sides of (3.18) one sees that S is even-dimensional. Every 2n-dimensional complex vector space admits such a map C which allows S to be given a *quaternionic structure* of a right H-module as follows. Take the real form of S — this a real 4n-dimensional space — and define the real endomorphisms J and K by

$$J(\phi) = \overline{C\phi} \qquad (3.19)$$

and

$$K(\phi) = \sqrt{-1}\,\phi \qquad (3.20)$$

It follows from (3.18) that the conditions (3.1) are satisfied. A complex endomorphism $f : S \to S$ is a quaternionic endomorphism if, and only if,

$$\bar{f}C = Cf. \qquad (3.21)$$

Example 3.1. Let T be a complex vector space. The complex space

$$S = T \otimes \bar{T}$$

has a natural real structure given by (3.15), where

$$C\,(\phi \otimes \bar{\psi}) = \bar{\psi} \otimes \phi \quad \text{for every } \phi, \psi \in S.$$

The real space S_+ consists of Hermitean tensors and is spanned by elements of the form $\phi \otimes \bar{\phi}$. In the notation of spinor calculus, the components of a generic element of S_+ constitute a Hermitean matrix $(\phi^{\mu\dot{\nu}})$,

$$\overline{\phi^{\mu\dot{\nu}}} = \phi^{\nu\dot{\mu}}$$

Example 3.2. Let T be a complex vector space. The complex space

$$S = T \oplus \bar{T}$$

has a natural real structure given by

$$C\,(\phi, \bar{\psi}) = (\bar{\psi}, \phi)$$

and a natural quaternionic structure given by

$$C'\,(\phi, \bar{\psi}) = (\bar{\psi}, -\phi).$$

3.5 Inner products in vector spaces

Consider a vector space V over the field $K = \mathbb{R}$ or $\mathbb{C}$ and a bilinear form

$$g : V \times V \to K$$

One says that g is *symmetric* if

$$g(u, v) = g(v, u) \quad \text{for every } u, v \in V.$$

If

$$g(u, v) = -g(v, u) \quad \text{for every } u, v \in V$$

then g is said to be *skew*. With any such form one associates a linear map from V to V*, also denoted by g,

$$g : V \to V^*$$

where

$$<g(u), v> = g(u, v) \quad \text{for every } u, v \in V \tag{3.22}$$

so that g is symmetric (resp., skew) if and only if, ${}^tg = g$ (resp. ${}^tg = -g$). The form g is said to be *non-degenerate* if the map $g : V \to V^*$ is bijective . If this is so, then the inverse map

$$g^{-1} : V^* \to V \tag{3.23}$$

defines a bilinear, non-degenerate form on V*

$$g^{-1}(\alpha, \beta) = <\alpha, g^{-1}(\beta)> \tag{3.24}$$

where $\alpha, \beta \in V^*$. The forms g and g^{-1} are simultaneously symmetric or skew.

A bilinear form g on V which is non-degenerate and either symmetric or skew is said to define an *inner product* on V; the pair (V, g) is an *inner product space.* A *scalar product* on V is a symmetric inner product. Two inner product spaces (V, g) and (W, h) are *isomorphic* if there is a linear map $f : V \to W$ such that

$$h(f(u), f(v)) = g(u, v) \tag{3.25}$$

for every $u, v \in V$. Clearly, g and h are then of the same symmetry type. If they are both symmetric, then f is called an *isometry*. An isometry of V onto itself is an *orthogonal* transformation.

The set O(g) of all isometries of (V, g) is the *orthogonal group* of (V, g). The *special orthogonal group* SO(g) consists of all isometries of (V, g) of determinant 1. If h is a skew product in W, then W is even-dimensional . The *symplectic group* Sp(h) is the group of all automorphisms of the inner product space (W, h). The following Proposition provides the classification of all finite-dimensional inner product spaces over $K = \mathbb{R}$ or $\mathbb{C}$:

Proposition 3.1. Every finite-dimensional inner product space of positive dimension is isomorphic to one of the following, where $k, l \in \mathbb{Z}$ and m is a positive integer:

(i) $(\mathbb{C}^m, g_m)$, where g_m is the scalar product

$$g_m(u, v) = u_1 v_1 + \dots + u_m v_m \quad (u_\alpha, v_\alpha \in \mathbb{C}, \quad \alpha = 1, \dots, m). \tag{3.26}$$

Notation: $O(g_m) = O(m, \mathbb{C})$.

(ii) $(\mathbb{C}^{2m}, h_m)$, where h_m is the skew inner product

$$h_m(u, v) = u_1 v_{m+1} - u_{m+1} v_1 + \dots + u_m v_{2m} - u_{2m} v_m \tag{3.27}$$

$(u_\alpha, v_\alpha \in \mathbb{C}, \quad \alpha = 1, \dots, 2m)$.

Notation: $Sp(h_m) = Sp(2m, \mathbb{C})$.

(iii) $(\mathbb{R}^m, g_{k,\ell})$, where $k + \ell = m$ and $g_{k,\ell}$ is the scalar product of *signature* (k, ℓ),

$$g_{k,\ell}(u, v) = u_1 v_1 + \dots + u_k v_k - u_{k+1} v_{k+1} - \dots - u_{k+\ell} v_{k+\ell} \tag{3.28}$$

$(u_\alpha, v_\alpha \in \mathbb{R}, \quad \alpha = 1, \dots, m)$.

Notation: $O(g_{k,\ell}) = O(k,\ell)$.

(iv) $(\mathbb{R}^{2m}, h'_m)$, where h'_m is as in (3.27) with $u_\alpha, v_\alpha \in \mathbb{R}$.

Notation: $Sp(h'_m) = Sp(2m, \mathbb{R})$.

The proof of the Proposition is classical and may be found in Bourbaki, Algebre, Ch. 9 (1959), for example.

If (V, g) and (W, h) are inner product spaces over K, then one can form their tensor product $(V \otimes W, g \otimes h)$ by putting

$$(g \otimes h)(v_1 \otimes w_1, v_2 \otimes w_2) = g(v_1, v_2)\, h(w_1, w_2) \tag{3.29}$$

where $v_1, v_2 \in V$ and $w_1, w_2 \in W$, and extending (3.29) bilinearly. It is clear that the tensor product of inner products has the following property

$$\text{symmetric} \otimes \text{symmetric} = \text{symmetric}, \tag{3.30}$$

$$\text{skew} \otimes \text{symmetric} = \text{skew}, \tag{3.31}$$

$$\text{skew} \otimes \text{skew} = \text{symmetric}. \tag{3.32}$$

The properties of the tensor product may be stated more precisely in terms of the signature and *index*, defined by

$$\text{index } g_{k,\ell} = k - \ell \in \mathbb{Z} \quad \text{for real spaces}$$

and

$$\text{index } g_m = m \pmod 2 \in \mathbb{Z}_2 \quad \text{for complex spaces}$$

with a scalar product. Then (3.30) may be sharpened to

$$g_m \otimes g_n = g_{mn}, \quad g_{k,\ell} \otimes g_{p,q} = g_{kp+\ell q,\, kq+\ell p} \tag{3.33}$$

so that for both $K = \mathbb{R}$ and $\mathbb{C}$

$$\text{index } (g \otimes g') = (\text{index } g).\,(\text{index } g') \tag{3.34}$$

and (3.32) may be replaced by the more precise statement

$$\text{index } (\text{skew} \otimes \text{skew}) = 0. \tag{3.35}$$

If (V, g) and (W, h) are inner product spaces over K and g and h are of the same symmetry type, then one can form their direct sum $(V \oplus W, g \oplus h)$ by defining

$$(g \oplus h)(v_1 \oplus w_1, v_2 \oplus w_2) = g(v_1, v_2) + h(w_1, w_2)$$

where $v_1, v_2 \in V$ and $w_1, w_2 \in W$. Clearly,

$$g_m \oplus g_n = g_{m+n}, \quad g_{k,\ell} \oplus g_{p,q} = g_{k+p,\, \ell+q}$$

and

$$\text{index } (g \oplus h) = \text{index } g + \text{index } h. \tag{3.36}$$

A space with a scalar product g of zero index is called *neutral*. Such a space is even-dimensional and can be characterized by the property of having totally null subspaces of half the dimension. Neutral spaces play an important role in the theory of Clifford algebras.

Let (V, g) be neutral of dimension 2m. It follows from Proposition 3.1. that there then exists a *null basis* in V such that the *quadratic form* g(u, u) is

$$u_1 u_{m+1} + u_2 u_{m+2} + \dots + u_m u_{2m} \tag{3.37}$$

where $u_\alpha \in K$ $(\alpha = 1, \dots, 2m)$ are the components of the vector u with respect to the null basis. If

$$W_i = \{u \in V \mid u_\alpha = 0 \text{ for } im + 1 \le \alpha \le im + m\}, \quad i = 0,1,$$

then

$$V = W_0 \oplus W_1$$

and the m-dimensional spaces W_0 and W_1 are totally null. (In the mathematical literature, the adjective *isotropic* is often used instead of null; we find that this can be somewhat confusing since isotropy has other connotations; incidentally, Cartan (1922) referred to null directions in a Lorentzian space-time as *optical*, but this name has not been accepted; see also Trautman (1985)).

The scalar product g in a real vector space V is said to be positive-definite if (V, g) is isomorphic to ($\mathbb{R}^n$, $g_{n,0}$) for some n; the scalar product g is negative-definite if -g is positive-definite; it is said to be definite if it is either positive- or negative-definite. Otherwise it is said to be indefinite.

If (V, g) is a real inner product space, then its *complexification* is the complex inner product space (W, h) such that $W = \mathbb{C} \otimes V$ and

$$h(u + iv, u' + iv') = g(u, u') - g(v, v') + i\,(g(u, v') + g(v, u')). \tag{3.38}$$

The symmetry of h is the same as that of g; one sometime writes

$$h = \mathbb{C} \otimes g$$

and refers to h as the complexification of g. The complexification of $g_{k,l}$ is equivalent to g_{k+l}.

Let g be a definite scalar product in an even-dimensional space V. The complexification h = $= \mathbb{C} \otimes g$ is neutral. Let N be a complex, totally null subspace of $W = \mathbb{C} \otimes V$ of maximal dimension. The complex conjugate subspace

$$P = \{u + iv \mid u - iv \in N; \quad u, v \in V\}$$

is also totally null of maximal dimension. If $u + iv \in N \cap P$, then $u - iv \in N$, and

$$h\,(u + iv, u - iv) = g\,(u, u) + g(v, v) = 0.$$

Since g is definite this implies $u = v = 0$ so that $N \cap P = \{0\}$ and there is the decomposition

$$W = N \oplus P.$$

Let n and p be the components of $u \in V$ in N and P, respectively. Since u is real, the vectors n and p are complex conjugate one to another. The vector i(n - p) is also real and

$$J(n+p) = i(n-p)$$

defines a linear map $J : V \to V$ such that $J^2 = -\mathrm{id}$ and $J \in O(g)$. Conversely, if J is an orthogonal complex structure in V, then

$$N = \{u - i J(u) \mid u \in V\}$$

is a maximal, totally null subspace of W. We have thus proved

Proposition 3.2. There is a natural one-to-one corespondence between the set of all orthogonal complex structures in an even-dimensional real space V with a definite scalar product and the set of all maximal totally null subspaces of $\mathbb{C} \otimes V$.

This has interesting global consequences. Consider, for example, the spheres S_{2m} of dimension 2m (m = 1, 2, ...). It is known (Steenrod 1951) that among them only S_2 and S_6 admit an almost complex structure. Therefore, for $m \neq 1$ and 3, the complexified tangent bundle of S_{2m} does not admit a smooth subbundle whose fibres are totally null of complex dimension m.

4. ALGEBRAS AND THEIR REPRESENTATIONS

4.1. Definitions

Our approach to spinors is based on Clifford algebras and their representations. It is convenient to present first the relevant notions in the general context of associative algebras. In particular, we show how the existence of a preferred inner product in the representation space of a simple algebra is related to that of an involutive antiautomorphism of the algebra.

An *algebra* over K is a vector space $\mathcal{A}$ over the field K, together with a composition law (product), i.e. a map $\mathcal{A} \times \mathcal{A} \to \mathcal{A}$, $(a, b) \to ab$ which is bilinear and distributive with respect to addition. The algebra is said to be *associative* if $a(bc) = (ab)c$ for every $a, b, c \in \mathcal{A}$. An element 1 of $\mathcal{A}$ is the *unity* if $1a = a1 = a$ for every $a \in \mathcal{A}$. The map $K \to \mathcal{A}$, given by $\lambda \to \lambda.1$, $\lambda \in K$, is injective and one identifies K with its image in $\mathcal{A}$; we write $\lambda \in \mathcal{A}$ instead of $\lambda.1$. A subset $\mathcal{G}$ of $\mathcal{A}$ is said to *generate* the algebra if every element of $\mathcal{A}$ can be expressed as a linear combination of products of elements of $\mathcal{G}$. For example, if $\mathcal{G}$ is a (linear) basis of $\mathcal{A}$, considered as vector space, then $\mathcal{G}$ generates $\mathcal{A}$.

Example 4.1. The set $\mathbb{C}$ of complex numbers, with standard multiplication, is a two-dimensional algebra over $\mathbb{R}$. This algebra is associative, commutative, has the unity 1 and is generated by the imaginary unit $\sqrt{-1}$. The pair $(1, \sqrt{-1})$ is a basis of $\mathbb{C}$.

Example 4.2. The set H of quaternions is a 4-dimensional associative algebra over $\mathbb{R}$. It is generated by $\mathcal{G} = \{i, j\}$, where $i^2 = j^2 = -1$ and $ij = -ji$. Any element of H is of the form

$$t + xi + yj + zk, \text{ where } k = ij \text{ and } t, x, y, z \in \mathbb{R}.$$

Example 4.3. Let S be a right module over the ring $L = \mathbb{R}$, $\mathbb{C}$ or H. The set $\mathrm{End}_L\, S$ of all endomorphisms of S (cf. § 3.2) is an algebra over $\mathbb{R}$ (the product is defined as the composition of endomorphisms). If $S = L^n$ then one writes

$$L(n) \text{ instead of } \mathrm{End}_L\, L^n.$$

The algebra $L(n)$ may be identified with the algebra of all n by n matrices with entries in L. Its dimension over $\mathbb{R}$ is n^2, $2n^2$, and $4n^2$ for $L = \mathbb{R}$, $\mathbb{C}$, and H, respectively. All these algebras are associative and with unity, but only $\mathbb{R}(1) = \mathbb{R}$ and $\mathbb{C}(1) = \mathbb{C}$ among them are commutative.

If $\mathcal{A}$ and $\mathcal{B}$ are algebras over K, then their direct sum $\mathcal{A} \oplus \mathcal{B}$ is the algebra defined as follows: its underlying vector space is the direct sum of the underlying vector spaces and multiplication in

$\mathcal{A} \oplus \mathcal{B}$ is given by

$$(a, b) \,.\, (a', b') = (aa', bb')$$

where a, a' $\in \mathcal{A}$ and b, b' $\in \mathcal{B}$. Similarly, the *tensor product* $\mathcal{A} \otimes \mathcal{B}$ is the algebra obtained by taking the tensor product of the underlying vector spaces and endowing it with the multiplication obtained by extending bilinearly the formula

$$(a \otimes b) \,.\, (a' \otimes b') = aa' \otimes bb'.$$

If $\mathcal{A}$ and $\mathcal{B}$ are algebras over K, then the map

$$f : \mathcal{A} \rightarrow \mathcal{B}$$

is said to be a *homomorphism* of algebras if it is K-linear and

$$f(ab) = f(a)\, f(b) \quad \text{for every } a, b \in \mathcal{A}.$$

If, moreover, both $\mathcal{A}$ and $\mathcal{B}$ are with unity and f(1) = 1, then f is said to be a homomorphism of algebras with unity. An invertible homomorphism is called an *isomorphism*.

The algebra $\mathbb{R}(2)$ is generated by the pair of matrices $\{\sigma, \tau\}$, where

$$\sigma = \begin{pmatrix} 0 & 1 \\ 1 & 0 \end{pmatrix}, \tau = \begin{pmatrix} 1 & 0 \\ 0 & -1 \end{pmatrix} \tag{4.1}$$

These matrices, together with the unit matrix I and

$$\varepsilon = \sigma\,\tau = \begin{pmatrix} 0 & -1 \\ 1 & 0 \end{pmatrix} \tag{4.2}$$

form a linear basis of $\mathbb{R}(2)$.

For any positive integers m and n, the algebras

$$\mathbb{R}(m) \otimes \mathbb{R}(n) \quad \text{and} \quad \mathbb{R}(mn) \tag{4.3}$$

as well as

$$L(\mathrm{n}) \text{ and } L \otimes \mathbb{R}(\mathrm{n}), \quad \text{where } L = \mathbb{R}, \mathbb{C} \text{ or } \mathrm{H}, \tag{4.4}$$

are isomorphic to each other.

It is useful also to know the tensor products $L_1 \otimes_{\mathbb{R}} L_2$, where L_1 and L_2 are $\mathbb{C}$ or H:

(i) there is an isomorphism $\mathbb{C} \otimes \mathbb{C} \to \mathbb{C} \oplus \mathbb{C}$ obtained by extending the following map between the generators of these algebras:

$$\sqrt{-1} \otimes 1 \to (\sqrt{-1}, \sqrt{-1}), \quad 1 \otimes \sqrt{-1} \to (\sqrt{-1}, -\sqrt{-1});$$

(ii) there is an isomorphism $\mathbb{C} \otimes \mathrm{H} \to \mathbb{C}(2)$ which on generators is

$$\sqrt{-1} \otimes 1 \to \sqrt{-1}\, I,$$
$$1 \otimes i \to -\sqrt{-1}\, \sigma,$$
$$1 \otimes j \to \varepsilon.$$

(iii) the isomorphism $\mathrm{H} \otimes \mathrm{H} \to \mathbb{R}(4)$ is obtained from

$$1 \otimes i \to \varepsilon \otimes \sigma, \quad 1 \otimes k \to \varepsilon \otimes \tau,$$
$$i \otimes 1 \to \sigma \otimes \varepsilon, \quad k \otimes 1 \to \tau \otimes \varepsilon.$$

The *centre* of an algebra $\mathcal{A}$ is the subalgebra $Z(\mathcal{A})$ consisting of all those elements which commute with all elements of $\mathcal{A}$. If $\mathcal{A}$ has a unit element, then $K \subset Z(\mathcal{A})$. If $Z(\mathcal{A}) = K$ then $\mathcal{A}$ is called a *central* algebra. The algebras $\mathbb{R}(n)$ and $\mathrm{H}(n)$ are central, but $\mathbb{C}(n)$, considered as an algebra over $\mathbb{R}$, is not: its centre is isomorphic to $\mathbb{C}$.

A *representation* of $\mathcal{A}$ in a vector space S over K is a homomorphism

$$\gamma : \mathcal{A} \to \mathrm{End}_K S.$$

A representation γ is said to be *faithful* if γ is injective.

A subspace T of S is called an *invariant space* of the representation γ if, for every $a \in \mathcal{A}$, one has $\gamma(a)\, T \subset T$. If T is an invariant space, then γ can be *reduced* to T, i.e. there is the representation

$$\gamma_T : \mathcal{A} \to \mathrm{End}_K T$$

defined by $\gamma_T(a)t = \gamma(a)t$ for $t \in T$ and $a \in \mathcal{A}$.

Consider now two representations of $\mathcal{A}$,

$$\gamma_i : \mathcal{A} \to \mathrm{End}_K\, S_i \qquad (i=1,2) \tag{4.5}$$

and assume that there is an *intertwining* transformation, i.e. a linear map

$$F : S_1 \to S_2 \tag{4.6}$$

such that

$$F\ \gamma_1 (a) = \gamma_2 (a)\ F \quad \text{for every } a \in \mathcal{A}. \tag{4.7}$$

The subspaces $\ker F = F^{-1}(0) \subset S_1$ and $F(S_1) \subset S_2$ are invariant spaces of γ_1 and γ_2, respectively. This observation leads to

Schur's Lemma : Let $F \neq 0$ intertwine the representations γ_1 and γ_2. If γ_1 is irreducible, then F is injective. If γ_2 is irreducible, then F is surjective. If both γ_1 and γ_2 are irreducible, then F is an isomorphism.

In the latter case, the representations γ_1 and γ_2 are said to be *equivalent*,

$$\gamma_1 \sim \gamma_2.$$

In particular, if $\gamma : \mathcal{A} \to \mathrm{End}_K\, S$ is irreducible, then the *commutant* of γ,

$$\{F \in \mathrm{End}_K\, S \mid F\ \gamma(a) = \gamma(a)\ F \ \text{ for every } a \in \mathcal{A}\}$$

is a *division algebra* : all its elements other than 0 are invertible. In particular, if $K = \mathbb{C}$ and S is finite-dimensional, then every $F \in \mathrm{End}_{\mathbb{C}}\, S$ has an eigenvalue, say

$$F\phi = \lambda\, \phi \quad \text{where } \lambda \in \mathbb{C} \ \text{ and } \ 0 \neq \phi \in S.$$

If F belongs to the commutant of the finite-dimensional, irreducible representation $\gamma : \mathcal{A} \to \mathrm{End}_{\mathbb{C}} S$, then $F - \lambda I$ also belongs to the commutant and $\phi \in \ker(F - \lambda I)$, therefore $F - \lambda I = 0$; this Corollary of Schur's Lemma is of frequent use.

Given two representations (4.1) of $\mathcal{A}$, one constructs their *direct sum*

$$\gamma_1 \oplus \gamma_2 : \mathcal{A} \to \mathrm{End}_K\, S_1 \oplus \mathrm{End}_K\, S_2 \tag{4.8}$$

and *tensor product*

$$\gamma_1 \otimes \gamma_2 : A \to \mathrm{End}_K (S_1 \otimes S_2) \tag{4.9}$$

as follows: for every $\phi_1 \in S_1$, $\phi_2 \in S_2$ and $a \in \mathcal{A}$ one puts

$$(\gamma_1 \oplus \gamma_2)(a)(\phi_1, \phi_2) = (\gamma_1(a)\phi_1, \gamma_2(a)\phi_2)$$

$$(\gamma_1 \otimes \gamma_2)(a)(\phi_1 \otimes \phi_2) = \gamma_1(a)\phi_1 \otimes \gamma_2(a)\phi_2.$$

If the space S of a representation γ decomposes into the direct sum, $S_1 \oplus S_2$, of two invariant spaces of γ, then γ is equivalent to the direct sum of the reductions γ_1 and γ_2 of γ to S_1 and S_2, respectively. One says that γ *decomposes* into the direct sum (4.8). It is clear that this notion generalizes to the case of more than two summands in a direct sum decomposition of S.

4.2 Simple algebras

Every algebra $\mathcal{A}$ over K admits the (left) *regular* representation

$$\rho : \mathcal{A} \to \mathrm{End}_K \mathcal{A} \tag{4.10}$$

defined by

$$\rho(a)b = ab, \quad \text{where } a, b \in \mathcal{A}.$$

The regular representation of an algebra with unity is faithful. An invariant space of the (left) regular representation is called a (left) *ideal* of the algebra. If $\mathcal{B} \subset \mathcal{A}$ is an ideal then there is the reduced representation

$$\rho_{\mathcal{B}} : \mathcal{A} \to \mathrm{End}_K \mathcal{B} \tag{4.11}$$

which is irreducible if, and only if, the ideal $\mathcal{B}$ is *minimal*, i.e. such that $\mathcal{B}$ contains no ideal other then $\{0\}$ and $\mathcal{B}$ itself.

Definition. An algebra over K is called *simple* if it admits a faithful and irreducible representation in a vector space over K. (An equivalent definition is: an algebra $\mathcal{A}$ is simple if $\{0\}$ and $\mathcal{A}$ are its only two-sided ideals).

For example, a division algebra is simple because its regular representation is faithful and irreducible.

Theorem 4.1. If an algebra is finite-dimensional and simple, then all its faithful irreducible representations are equivalent.

The importance of this theorem justifies sketching its *proof.* A non-zero finite-dimensional algebra $\mathcal{A}$ contains a non-zero minimal left ideal $\mathcal{B}$. Let γ be a faithful irreducible representation of $\mathcal{A}$ in a vector space S. Since γ is faithful, there exist elements $\phi_o \in S$ and $b_o \in \mathcal{B}$ such that $\gamma(b_o)\phi_o \neq 0$. Let $F : B \rightarrow S$ be defined by $F(b) = \gamma(b)\phi_o$. The linear map F intertwines the irreducible representations γ and $\rho_{\mathcal{B}}$. Since $F(\mathcal{B})$ contains non-zero elements, the map F is an isomorphism. Therefore, the representation γ— and thus every faithful irreducible representation of $\mathcal{A}$— is equivalent to $\rho_{\mathcal{B}}$.

It follows from the Theorem that the faithful irreducible representation of a finite-dimensional simple algebra can be realized by left multiplication on a left minimal ideal. In other words, such an ideal is the carrier space of the representation. This observation is at the basis of the "algebraic theory of spinors" (Chevalley 1954).

The tensor product of simple algebras need not be simple (example: $\mathbb{C}$ is simple, but $\mathbb{C} \otimes_{\mathbb{R}} \mathbb{C}$ is not). However, *if* $\mathcal{A}$ *is central simple and* $\mathcal{A}'$ *is simple, then* $\mathcal{A} \otimes \mathcal{A}'$ *is simple*; for a proof, see J.P. Serre (1955). If an algebra is the direct sum of simple algebras, then it is said to be *semi-simple.*

The commutant of a faithful irreducible representation of a central simple algebra over $\mathbb{R}$ need not be trivial. For example, the algebra H admits a representation γ in $\mathbb{R}^4 = \mathbb{R}^2 \otimes \mathbb{R}^2$ such that $\gamma(i) = \varepsilon \otimes \sigma$ and $\gamma(k) = \varepsilon \otimes \tau$. This representation is faithful irreducible and its commutant is the division algebra isomorphic to H, generated by the pair $\{\sigma \otimes \varepsilon, \tau \otimes \varepsilon\}$. From now on, through the end of this chapter, we consider only finite-dimensional associative algebras with unity and their representations in finite-dimensional vector spaces.

4.3. Antiautomorphisms and Inner products

A linear invertible map $\beta : \mathcal{A} \rightarrow \mathcal{A}$ is said to be an *antiautomorphism* of the algebra $\mathcal{A}$ if

$$\beta(1) = 1 \text{ and } \beta(ab) = \beta(b)\,\beta(a) \tag{4.12}$$

for every a, b $\in \mathcal{A}$. It is said to be *involutive* if

$$\beta^2 = \text{id}. \tag{4.13}$$

If β is an antiautomorphism of $\mathcal{A}$ and γ is a representation in S, then there is another representation

$$\check{\gamma} : \mathcal{A} \to \text{End } S^*$$

defined by

$$\check{\gamma}(a) = {}^t\gamma(\beta(a)).$$

If γ is a faithful irreducible representation of a simple algebra $\mathcal{A}$ with an antiautomorphism β, then $\check{\gamma}$ is equivalent to γ : there thus exists an isomorphism

$$B : S \to S^*$$

such that

$$ {}^t\gamma(\beta(a))\, B = B\, \gamma(a) \tag{4.14}$$

for every $a \in \mathcal{A}$. The isomorphism B defines an inner product on S, also denoted by B,

$$B(\phi, \psi) = \langle B(\phi), \psi \rangle$$

where $\phi, \psi \in S$. For every $a \in \mathcal{A}$, by virtue of (4.14), one has

$$B(\gamma(a)\,\phi, \gamma(a)\,\psi) = B(\phi, \gamma(\beta(a)\,a)\,\psi) \tag{4.15}$$

and also

$$\gamma(a)\, B^{-1}\,({}^tB) = B^{-1}\,({}^tB)\, \gamma(\beta^2(a)).$$

Therefore, if β is involutive, then $B^{-1}\,({}^tB)$ belongs to the commutant of γ. If $\mathcal{A}$ and S are over $\mathbb{C}$ then, by the Corollary of Schur's Lemma, ${}^tB$ is proportional to B and, since ${}^{tt}B = B$,

$$\text{either } {}^tB = B \text{ or } {}^tB = -B. \tag{4.16}$$

If β is an antiautomorphism of the algebra $\mathcal{A}$, then one can associate with the pair $(\mathcal{A}, \beta)$ the group of all elements of the algebra such that the *norm* $N(a) = \beta(a)\, a$ is an invertible element of K. Since the regular representation is faithful, one has $\beta(a)\, a = a\, \beta(a)$ for any element a of the group; the unity of $\mathcal{A}$ is the neutral element of the group. If $\lambda \in K^{\bullet} = K - \{0\}$ and a belongs to the group, then so does λa and $N(\lambda\, a) = \lambda^2\, N(a)$. We define the *Wall group* $\mathcal{G}$ of the pair $(\mathcal{A}, \beta)$ as

$$\mathcal{G} = \{a \in \mathcal{A} : N(a) \in K^{\bullet}/K^{\bullet 2}\} \tag{4.17}$$

where $K^{\bullet 2}$ is the subgroup of squares of $K^{\bullet}$, i.e.

$$K^\bullet / K^{\bullet 2} = \begin{cases} \{1\} & \text{for } K = \mathbb{C} \\ \{1, -1\} & \text{for } K = \mathbb{R} \end{cases}$$

The group

$$\mathcal{G}_0 = \{a \in \mathcal{A} : N(a) = 1\} \tag{4.18}$$

coincides with $\mathcal{G}$ for $K = \mathbb{C}$, but may be a proper subgroup of $\mathcal{G}$ for $K = \mathbb{R}$. We can now formulate

Proposition 4.1. Let $\gamma : \mathcal{A} \to \mathrm{End}_K\, S$ be a faithful irreducible representation of a finite-dimensional associative simple algebra endowed with an antiautomorphism β. There then exists an inner product B on S which satisfies

$$B(\gamma(a)\,\phi, \gamma(a)\,\psi) = N(a)\, B\,(\phi, \psi) \tag{4.19}$$

for every $\phi, \psi \in S$ and a in the Wall group $\mathcal{G}$. Moreover, if the group $\mathcal{G}$ generates $\mathcal{A}$ and B' is another inner product with the property (4.19), then $B^{-1}B'$ is in the commutant of γ.

In particular if both $\mathcal{A}$ and S are over $\mathbb{C}$, then the last part of the Proposition implies that the inner product on S, invariant with respect to $\mathcal{G}$, is essentially unique: it is defined up to a numerical factor.

Example 4.4 Let $\mathcal{A} = L(n)$, where $L = \mathbb{R}$, $\mathbb{C}$ or H and n = 1, 2, ... For any matrix $a \in L(n)$, let $\beta(a) = a^\dagger$ be its Hermitean conjugate matrix; for $L = \mathbb{R}$ the matrix $a^\dagger$ is the transpose of a. The map $\beta : L(n) \to L(n)$ is an involutive antiautomorphism of the algebra $L(n)$ over $\mathbb{R}$ and

$$\mathcal{G}_0 = \begin{cases} O(n) \\ U(n) \\ Sp(n) \end{cases} \quad \text{for } L = \begin{cases} \mathbb{R} \\ \mathbb{C} \\ H \end{cases}$$

Let $S_L(n) = \mathbb{R}^n$, $\mathbb{R}^2 \otimes \mathbb{R}^n$ or $\mathbb{R}^2 \otimes \mathbb{R}^2 \otimes \mathbb{R}^n$ depending on whether $L = \mathbb{R}$, $\mathbb{C}$ or H. The algebra $L(n)$ has a faithful irreducible representation γ in $S_L(n)$ which can be described as follows:

(i) if $L = \mathbb{R}$, then $\gamma(a) = a$;
(ii) if $L = \mathbb{C}$, then $\gamma(p+\sqrt{-1}\,q) = I \otimes p + \varepsilon \otimes q$, where $p, q \in \mathbb{R}(n)$;
(iii) if $L = H$, then $\gamma(p) = I \otimes I \otimes p$, $\gamma(i) = \varepsilon \otimes \sigma \otimes I$ and $\gamma(k) = \varepsilon \otimes \tau \otimes I$, where I denotes the unit matrix of appropriate dimension.

The standard, positive-definite scalar product on $S_L(n)$ is invariant with respect to $\mathcal{G}_0$ in every case.

The standard, positive-definite scalar product on $S_L(n)$ is invariant with respect to $\mathcal{G}_0$ in every case.

An involutive antiautomorphism β of a simple algebra $\mathcal{A}$ defines a bilinear symmetric function

$$h : \mathcal{A} \times \mathcal{A} \to K$$

given by

$$h(a,b) = \operatorname{Tr} \gamma (\beta (a)\, b) \tag{4.20}$$

where γ is a faithful irreducible representation of $\mathcal{A}$ in S. Indeed, by virtue of (4.14) one has

$$\operatorname{Tr} \gamma (\beta (a)) = \operatorname{Tr} \gamma (a)$$

and therefore

$$\operatorname{Tr} \gamma (\beta (b)\, a) = \operatorname{Tr} \gamma (\beta (a)\, \beta^2 (b)) = \operatorname{Tr} \gamma (\beta (a)\, b)$$

so that h is indeed symmetric.

<u>Proposition 4.2.</u> Let the simple algebra $\mathcal{A}$ be endowed with the bilinear form h defined by (4.20), where β is an involutive antiautomorphism. A faithful irreducible representation $\gamma : \mathcal{A} \to \operatorname{End}_K S$ is an isometry of $(\mathcal{A}, h)$ on the image $\gamma(\mathcal{A}) \subset \operatorname{End}_K S$ equipped with the restriction of the scalar product $B \otimes B^{-1}$.

To prove the Proposition, recall (§ 3.5) that if S is a vector space with an inner product B, then $S \otimes S^*$ can be given the scalar product $B \otimes B^{-1}$ in such a way that

$$(B \otimes B^{-1}) (\phi \otimes \phi', \psi \otimes \psi') = \langle B (\phi), \psi \rangle \langle \phi', B^{-1} (\psi') \rangle \tag{4.21}$$

where $\phi, \psi \in S$ and $\phi', \psi' \in S^*$. The identification of End S with $S \otimes S^*$ defined by

$$(\phi \otimes \phi') (\psi) = \langle \phi', \psi \rangle\, \phi, \quad \text{where } \phi, \psi \in S \text{ and } \phi' \in S^*, \tag{4.22}$$

leads to

$$\operatorname{Tr} (\phi \otimes \phi') = \langle \phi', \phi \rangle \quad \text{and} \quad {}^t(\phi \otimes \phi') = \phi' \otimes \phi. \tag{4.23}$$

For every $a \in \mathcal{A}$ there exist sequences $(\phi_1, \ldots, \phi_m)$ and $(\phi_1', \ldots, \phi'_m)$ of elements of S and S*, respectively, such that

$$\gamma(a) = \Sigma\ \phi_i \otimes \phi_i' \tag{4.24}$$

To evaluate h(a, b) by means of (4.20) we can write $\gamma(a) = \Sigma\, \phi \otimes \phi'$ and $\gamma(b) = \Sigma\, \psi \otimes \psi'$, where we have neglected the indices labelling the vectors in the decomposition (4.24) and in a similar decomposition of $\gamma(b)$. Since, by virtue of (4.14) and (4.22),

$$^t\gamma(\beta\,(a)) = B\,\gamma\,(a)\,B^{-1} = \Sigma\, B(\phi) \otimes {}^tB^{-1}\,(\phi')$$

we obtain

$$\begin{aligned} h(a, b) &= Tr\,{}^t\gamma(\beta\,(a)\,b) = \Sigma\,\Sigma\, Tr\,(\psi' \otimes \psi)\,(B(\phi) \otimes {}^tB^{-1}\,(\phi')) \\ &= \Sigma\,\Sigma <B\,(\phi), \psi> <\phi', B^{-1}\,(\psi')> \end{aligned}$$

By comparing this with (4.20) one gets

$$h = (B \otimes B^{-1}) \circ \gamma \qquad (4.25)$$

as asserted in the Proposition.

It is sometimes of interest to know how the inner product behaves under a change of representation. Let B_1 and B_2 be the inner products in the carrier spaces S_1 and S_2 of two faithful irreducible representations (4.5) of $\mathcal{A}$. If F intertwines the representations, then

$$B_1 = {}^tF\, B_2\, F.$$

4.4 Real algebras

We have so far considered algebras over the field *K* of either real or complex numbers and their representations in vector spaces over the *same* field *K*. If $\mathcal{A}$ is a *real* algebra, i.e. an algebra over $\mathbb{R}$, then, besides its representations in real vector spaces, one can consider *complex representations* of $\mathcal{A}$. i.e. representations of $\mathcal{A}$ in complex vector spaces. More precisely

$$\gamma : \mathcal{A} \to \mathrm{End}_{\mathbb{C}} S$$

is a complex representation of the real algebra $\mathcal{A}$ in the complex vector space S if

$$\gamma\,(a)\,(\lambda\phi + \mu\psi) = \lambda\,\gamma\,(a)\,\phi + \mu\,\gamma\,(a)\,\psi$$

for every $\lambda, \mu \in \mathbb{C}$ and $\phi, \psi \in S$, as well as

$$\gamma(\rho a + \sigma b) = \rho\,\gamma(a) + \sigma\,\gamma(b)$$
$$\gamma(ab) = \gamma(a)\,\gamma(b)$$

for every $\rho, \sigma \in \mathbb{R}$ and $a, b \in \mathcal{A}$. Since a complex vector space may be considered also as a real vector space (of double real dimension), any complex representation γ defines a real one, called the *real form* of γ. For example, the real algebra H has a complex representation γ in $S = \mathbb{C}^2$ determined by

$$\gamma(i) = -\sqrt{-1}\,\sigma \text{ and } \gamma(k) = -\sqrt{-1}\,\tau. \tag{4.26}$$

Its real form γ' is a representation in $\mathbb{R}^4$ obtained from

$$\gamma'(i) = \varepsilon \otimes \sigma,\ \ \gamma'(k) = \varepsilon \otimes \tau, \tag{4.27}$$

in agreement with Example 4.4.

Let

$$\gamma_i : \mathcal{A} \to \mathrm{End}_{\mathbb{C}} S_i \quad (i = 1, 2)$$

be two complex representations of a real algebra $\mathcal{A}$. They are said to be complex-equivalent if there is a $\mathbb{C}$-linear isomorphism $F : S_1 \to S_2$ intertwining γ_1 and γ_2, i.e. such that (4.7) holds.

Complex equivalence is a stronger relation than the equivalence of the real forms of complex representations. For example, the algebra of complex numbers, considered as a real algebra, admits two complex-inequivalent representations γ_1 and γ_2 in $S = \mathbb{C}$,

$$\gamma_1(x + iy) = x + iy,\ \ \gamma_2(x + iy) = x - iy,\ \ \text{where } x, y \in \mathbb{R} \text{ and } i = \sqrt{-1}.$$

Their real forms are representations in $\mathbb{R}^2$ which are equivalent because $\mathbb{C}$ is simple; explicitly, since

$$\gamma_1'(x + iy) = x.I + y\,\varepsilon,\ \ \gamma_2'(x + iy) = x.I - y\,\varepsilon,$$

we have

$$\gamma_1'(x + iy)\,\tau = \tau\,\gamma_2'(x + iy).$$

If γ is a complex representation of the real algebra $\mathcal{A}$ in S, then the *complex conjugate representation* $\bar{\gamma}$ is a complex rerpresentation of the same algebra in the space $\bar{S}$, complex conjugate to S,

$$\bar{\gamma} : \mathcal{A} \to \mathrm{End}_{\mathbb{C}} \bar{S}$$

given by

$$\bar{\gamma}(a) = \overline{\gamma(a)}.$$

If $\mathcal{A}$ is a real algebra, then its *complexification* $\mathbb{C} \otimes \mathcal{A}$ has the structure of an algebra over $\mathbb{C}$. A complex representation γ of $\mathcal{A}$ extends to a representation of the complexification in the same vector space. We use the same letter γ to denote the extended representation, thus

$$\gamma(a + ib) = \gamma(a) + i\gamma(b)$$

where $i = \sqrt{-1}$ and $a, b \in \mathcal{A}$.

<u>Theorem 4.2.</u> Let $\gamma : \mathcal{A} \to \mathrm{End}_{\mathbb{C}} S$ be a complex, faithful and irreducible representation of a central simple algebra $\mathcal{A}$ over $\mathbb{R}$. There then exists a $\mathbb{C}$-linear isomorphism

$$C : S \to \bar{S} \tag{4.28}$$

which intertwines the representations γ and $\bar{\gamma}$,

$$\bar{\gamma}(a)\, C = C\, \gamma(a), \quad a \in \mathcal{A}, \tag{4.29}$$

and is such that

$$\text{either} \quad \bar{C}C = \mathrm{id} \quad \text{or} \quad \bar{C}C = -\mathrm{id}. \tag{4.30}$$

Moreover, if $\mathcal{A}$ has an involutive antiautomorphism β, then the isomorphism $B : S \to S^*$ intertwining γ and $\check{\gamma}$ can be chosen so that the linear map

$$A = \bar{B}\, C : S \to \bar{S}^* \tag{4.31}$$

is Hermitean, i.e.,

$${}^{t}C\, B^{\dagger} = B\, \bar{C}. \tag{4.32}$$

<u>Proof.</u> Since $\mathcal{A}$ is central simple over $\mathbb{R}$, the complexification $\mathbb{C} \otimes \mathcal{A}$ is central simple over $\mathbb{C}$ and the complex extension of γ is a faithful irreducible representation of $\mathbb{C} \otimes \mathcal{A}$ in S. The same is true of $\bar{\gamma}$; therefore, the complex extensions of γ and $\bar{\gamma}$ are equivalent; let C be the intertwining isomorphism. By complex conjugation of (4.29) we obtain $\gamma(a)\, \bar{C} = \bar{C}\, \bar{\gamma}(a)$; therefore $\bar{C}C$ is in the

commutant of γ, and also of its complex extension. By the Corollary of Schur's Lemma, the endomorphism $\bar{C}C$ of S is a multiple of the identity. If C is replaced by λC, where $\lambda \in \mathbb{C}^*$, then $\bar{C}C$ is replaced by $|\lambda|^2 \bar{C}C$. By choosing λ appropriately, one achieves (4.30).

From (4.14) and (4.29) it follows that $B\,\bar{C}\,({}^t C\, B^\dagger)^{-1}$ is in the commutant of γ; therefore ${}^t B\,\bar{C} = \mu\,{}^t C\, B^\dagger$ for some $\mu \in \mathbb{C}$. By Hermitean conjugation of the last equation one obtains ${}^t C\, B^\dagger = \bar{\mu}\, B\, \bar{C}$; therefore $|\mu| = 1$ and one can satisfy (4.32) by replacing B with $\sqrt{\mu}\, B$.

The linear map A defines a *Hermitean* form on S, also denoted by A,

$$A(\phi, \psi) = \langle \bar{A}\,(\bar{\phi}), \psi \rangle, \quad \text{where } \phi, \psi \in S. \tag{4.33}$$

For every element a of the Wall group $\mathcal{G}$ we have

$$A\,(\gamma\,(a)\,\phi, \gamma\,(a)\,\psi) = N(a)\; A\,(\phi, \psi) \tag{4.34}$$

Example 4.5. Let $\gamma : H \to \mathrm{End}_{\mathbb{C}}\, \mathbb{C}^2$ be the complex representation of the algebra of quaternions defined by (4.26). The conjugation of quaternions defines an involutive antiautomorphism β of H. The algebra H is central simple and, therefore, $\bar{\gamma} \sim \gamma \sim \check{\gamma}$. One easily sees that $C = \varepsilon = B$ provide the intertwining isomorphisms. On the other hand, the simple algebra $\mathbb{C}(2)$ is *not central* (as an algebra over $\mathbb{R}$), and its complexification $\mathbb{C} \otimes_{\mathbb{R}} \mathbb{C}(2) = 2\mathbb{C}(2)$ is not simple. The identity representation γ of $\mathbb{C}(2)$ in $S = \mathbb{C}^2$ is faithful irreducible, but $\bar{\gamma}$ is not complex equivalent to γ. (In the calculus of two-component spinors these observations correspond to the following statement: "dotted" and "undotted" spinors provide inequivalent representations of SL(2, $\mathbb{C}$); these representations become equivalent when the group is restricted to SU(2) = Sp(1)).

Assume now that the conditions stated in Proposition 4.3 are satisfied and

$$\bar{C}C = \mathrm{id}.$$

There is then the direct sum decomposition (3.17) of S into two real vector spaces S^+ and S^- such that

$$\phi \in S^\pm \;\text{ iff }\; \bar{\phi} = \pm C\phi.$$

The map $K : S^+ \to S^-$ given by $K(\phi) = \sqrt{-1}\,\phi$ is a real isomorphism. The subspaces S^+ and S^- are invariant with respect to γ and the complex representation γ reduces to two real ones

$$\gamma^\pm : \mathcal{A} \to \mathrm{End}_{\mathbb{R}}\, S^\pm.$$

The representations γ^+ and γ^- are equivalent,

$$\gamma^- K = K \gamma^+.$$

By an abuse of language one often says that γ is a *real representation.*

Similarly, if

$$\bar{C}C = -\mathrm{id},$$

then S has the structure of a right H-module, defined by (3.19,20) and the representation γ is *quaternionic* in the sense of consisting of endomorphisms of the quaternionic structure of S.

4.5. Graded algebras

An algebra $\mathcal{A}$ over K is said to be $\mathbb{Z}_2$-*graded* or simply *graded,* if there is a decomposition of $\mathcal{A}$ into a direct sum of two vector spaces

$$\mathcal{A} = \mathcal{A}_0 \oplus \mathcal{A}_1$$

such that

$$\mathcal{A}_p \mathcal{A}_q \subset \mathcal{A}_{p+q}$$

where the sum p+q is taken mod 2, i.e. $1 + 1 \equiv 0$. The elements of $\mathcal{A}_0$ (resp., $\mathcal{A}_1$) are said to be *even* (resp., *odd*). Sometimes one writes deg a = p if $a \in \mathcal{A}_p$. To describe a graded algebra, it is often sufficient to indicate the embedding

$$\mathcal{A}_0 \to \mathcal{A}. \qquad (4.35)$$

A *graded derivation* of degree $p \in \mathbb{Z}_2$ is a linear map $d : \mathcal{A} \to \mathcal{A}$ such that

$$d\,\mathcal{A}_q \subset \mathcal{A}_{p+q}$$

and

$$d\,(ab) = (da)b + (-1)^{pq}\, a\, db$$

for every $a \in \mathcal{A}_q$ and $b \in \mathcal{A}$. A graded derivation of degree 0 is called simply a derivation; a graded derivation of degree 1 is an *antiderivation.* A *graded Lie algebra* (or super Lie algebra) is a graded algebra

$$\mathcal{L} = \mathcal{L}_0 \oplus \mathcal{L}_1$$

with a product [,] which is "superanticommutative",

$$[a, b] = -(-1)^{pq} [b, a] \quad \text{for every } a \in \mathcal{A}_p \text{ and } b \in \mathcal{A}_q,$$

and such that the map

$$\text{ad}(a) : \mathcal{L} \rightarrow \mathcal{L}, \text{ where } \langle \text{ad}(a), b \rangle = [a, b],$$

is a graded derivation of degree p = deg a. With any graded algebra $\mathcal{A}$ one associates the graded Lie algebra Der $\mathcal{A}$ of derivations by defining

$$[d_p, d_q] = d_p \circ d_q - (-1)^{pq} d_q \circ d_p,$$

where d_p and d_q are graded derivations of degree p and q, respectively. If the graded algebra $\mathcal{A}$ is associative, then one can form the graded Lie algebra $\mathcal{L}(\mathcal{A})$: its underlying vector space and grading are those of $\mathcal{A}$ and the product is given by the graded (super) commutator,

$$[a, b] = ab - (-1)^{pq} ba, \quad \text{where } a \in \mathcal{A}_p \text{ and } b \in \mathcal{A}_q. \tag{4.36}$$

The *supercentre* SZ($\mathcal{A}$) of a graded associative algebra $\mathcal{A}$ is the set

$$\{a \in \mathcal{A} \mid [a, b] = 0 \quad \text{for every } b \in \mathcal{A}\}$$

where the bracket is defined by (4.36). If SZ($\mathcal{A}$) = $\mathcal{A}$ then $\mathcal{A}$ is said to be *supercommutative*. If $\mathcal{A}$ has a unity 1, then $1 \in \mathcal{A}_0$; in this case, $\mathcal{A}$ is said to be *supercentral* if its supercentre coincides with the number field $K \subset \mathcal{A}_0$.

Example 4.6. The algebra $\mathbb{C}$ is a graded associative algebra over $\mathbb{R}$. The grading is defined by declaring pure imaginary numbers to be odd. This algebra is supercentral, but not supercommutative, $[\sqrt{-1}, \sqrt{-1}] = -2$. It is not central over $\mathbb{R}$.

Example 4.7. The simple algebra $\mathbb{R}(2)$ can be given two essentially different gradings:
(a) by declaring the matrices σ and τ to be odd, one obtains an even subalgebra isomorphic to $\mathbb{C}$, thus (4.35) is now $\mathbb{C} \rightarrow \mathbb{R}(2)$;
(b) if the matrices σ and ε are odd, then the even subalgebra is spanned by I and τ: it is isomorphic to $2\mathbb{R}$.

In either of the gradings the algebra is supercentral.

There is a number of standard constructions that lead to graded algebras. In the sequel, $\mathcal{A}$ and $\mathcal{B}$ denote a graded and an ungraded algebra, respectively.

I. The *double* of $\mathcal{B}$ is the graded algebra

$$\mathcal{B} \to 2\mathcal{B} \tag{4.37}$$

where

$$2\mathcal{B} = \mathcal{B} \oplus \mathcal{B}$$

and elements of the form (b, -b), $b \in \mathcal{B}$, are odd.

II. The *tensor product* $\mathcal{A} \otimes \mathcal{B}$ has a standard grading,

$$(\mathcal{A} \otimes \mathcal{B})_p = \mathcal{A}_p \otimes \mathcal{B}, \ p = 0, 1. \tag{4.38}$$

III. If both $\mathcal{A}$ and $\mathcal{A}'$ are graded, then their *graded tensor product* $\mathcal{A} \otimes \mathcal{A}'$ has $\mathcal{A} \otimes \mathcal{A}'$ as its underlying vector space,

$$\deg (a \otimes a') \equiv \deg a + \deg a' \pmod 2$$

and

$$(a \otimes a'_p) \cdot (b_q \otimes b') = (-1)^{pq}\, ab_q \otimes a'_p b' \tag{4.39}$$

for every $a \in \mathcal{A}$, $b_q \in \mathcal{A}_q$, $a'_p \in \mathcal{A}'_p$ and $b' \in \mathcal{A}'$.

Example 4.8. Let $\mathbb{L}$ be the graded algebra over $\mathbb{R}$ described in Example 4.6. The graded tensor product

$$\mathbb{C} \otimes_{\mathbb{R}} \mathbb{C} \text{ is isomorphic to } H \tag{4.40}$$

where the algebra H is graded by declaring j and k to be odd. Indeed, a (graded) isomorphism is obtained by extending the map $\sqrt{-1} \otimes 1 \to j$, $1 \otimes \sqrt{-1} \to k$.

IV. If $\mathcal{A}$ is graded, then the opposite algebra $\mathcal{A}^{opp}$ has the same underlying vector space structure and grading as $\mathcal{A}$, and an opposite product x defined by

$$a \times b = (-1)^{pq}\, ba$$

for $a \in \mathcal{A}_p$ and $b \in \mathcal{A}_q$. Therefore, $\mathcal{A}$ is supercommutative iff the identity map $\mathcal{A} \to \mathcal{A}^{opp}$ is an isomorphism of graded algebras.

Example 4.9. Consider the double of $\mathbb{R}$, i.e. the graded algebra $\mathcal{A}_0 = \mathbb{R} \to 2\mathbb{R} = \mathcal{A}$. The algebra $\mathbb{R} \to \mathbb{C}$ is opposite to $\mathbb{R} \to 2\mathbb{R}$: since the odd part of $\mathcal{A}$ is generated by $(1, -1)$ and $(1, -1) \times (1, -1) = -(1, 1)$, the linear map

$$(\lambda, \mu) \to \frac{1}{2} \; (\lambda + \mu + \sqrt{-1}\; (\lambda - \mu)), \qquad \text{where } \lambda, \mu \in \mathbb{R},$$

is a graded isomorphism of $\mathcal{A}^{opp}$ on $\mathbb{C}$.

Example 4.10. Let $\mathcal{A}_0 = 2\mathbb{R} \to \mathbb{R}(2) = \mathcal{A}$ be the graded algebra of Example 4.7(b). The linear map $f : \mathcal{A} \to \mathcal{A}^{opp}$ defined by $f(I) = I$, $f(\tau) = -\tau$, $f(\varepsilon) = \sigma$ and $f(\sigma) = \varepsilon$ is an isomorphism of graded algebras.

Example 4.11. Let $\mathcal{A}_0 = \mathbb{C} \to \mathbb{R}(2) = \mathcal{A}$ be the graded algebra of Example 4.7(a) and let $\mathcal{B}_0 = \mathbb{C} \to H = \mathcal{B}$ be the graded algebra described in Example 4.8. The linear map $f : \mathcal{B} \to \mathcal{A}^{opp}$ defined by $f(1) = I$, $f(i) = -\varepsilon$, $f(j) = \sigma$ and $f(k) = \tau$ is an isomorphism of graded algebras.

5. GENERAL PROPERTIES OF CLIFFORD ALGEBRAS

Much of mathematics and physics involves the idea of *linearization.* For example, the tensor product of vector spaces is introduced in order to represent any bilinear map as a linear map composed with the canonical bilinear map. The derivative of a differentiable map is, in a well-known sense, its linear part. Clifford algebras can also be so viewed: they allow a *linearization of quadratic forms*. Preceded by the discovery of quaternions by Hamilton, these algebras were defined by Clifford who found their general structure. An anonymous correspondent, signing in 1959 his letter with the name of R. Lipschitz, reminds the scientific community that it was Lipschitz who for the first time used Clifford algebras and groups to represent orthogonal transformations in an n-dimensional space. In a sense, Lipschitz's memoir (1886) is the first publication on the relation between spinor and orthogonal groups.

There are many excellent and exhaustive descriptions of Clifford algebras (Chevalley 1954, Artin 1957, Bourbaki 1959, Atiyah, Bott and Shapiro 1964, Porteous 1981). In this chapter, we only collect the basic definitions and results. More emphasis is put here than elsewhere on the invariant bilinear forms on spinor spaces.

All vector spaces and algebras considered here are over the field K of real or complex numbers. The vector spaces are finite-dimensional and algebras are associative with unity.

5.1. Definition and general properties of Clifford algebras

Let g denote a scalar product in a vector space V over K and let $\mathcal{A}$ be an algebra over the same field. A Clifford map of (V, g) into $\mathcal{A}$ is a linear map

$$f : V \to \mathcal{A}$$

such that, for any $u \in V$,

$$f(u)^2 = g(u, u). \tag{5.1}$$

For example, if $K = \mathbb{R}$, $V = \mathbb{R}^2$ with the standard scalar product $g = g_{2,0}$ and $\mathcal{A} = \mathbb{R}(2)$, then $f : V \to \mathcal{A}$ given by

$$f(x, y) = \begin{pmatrix} x & y \\ y & -x \end{pmatrix}$$

is a Clifford map.

Let

$$\mathcal{T}(V) = \bigoplus_{i=0}^{\infty} \otimes^{i} V, \tag{5.2}$$

be the tensor algebra of V and let $\mathcal{I}(g)$ be its two-sided ideal generated by all elements of the form

$$u \otimes u - g(u, u), \quad \text{where } u \in V.$$

The factor algebra

$$\mathcal{T}(V) \,/\, \mathcal{I}(g)$$

is called the Clifford algebra of (V, g) and denoted by $\mathcal{Cl}(g)$. The canonical map

$$\kappa : \mathcal{T}(V) \to \mathcal{Cl}(g) \tag{5.3}$$

restricted to the subspace V of $\mathcal{T}(V)$ is a Clifford map. Indeed, κ is linear and, if multiplication in $\mathcal{Cl}(g)$ is denoted by a dot (often neglected), so that

$$\kappa\,(a \otimes b) = \kappa\,(a)\,.\,\kappa\,(b) \quad \text{for every } a, b \in \mathcal{T}(V),$$

then

$$\kappa\,(u)\,.\,\kappa\,(u) = \kappa\,(u \otimes u) = \kappa\,(g\,(u, u)) = g\,(u, u) \tag{5.4}$$

Theorem 5.1. The Clifford map $\kappa : V \to \mathcal{Cl}(g)$ is universal, i.e. if $f : V \to \mathcal{A}$ is a Clifford map of (V, g) into $\mathcal{A}$, then there exists exactly one homomorphism of algebras with unity $\tilde{f} : \mathcal{Cl}(g) \to \mathcal{A}$ such that $f = \tilde{f} \circ \kappa$.

This is easily proved by extending f to a homomorphism $\hat{f}$ of $\mathcal{T}(V)$ into $\mathcal{A}$ and noting that $\hat{f}$ vanishes on $\mathcal{I}(g)$.

Our definitions follow the terminology accepted in mathematics. It should be noted, however, that the interesting object is not the algebra $\mathcal{Cl}(g)$ by itself, but the Clifford map

$$\kappa : V \to C\ell(g). \tag{5.5}$$

It contains information not only about the algebra, but also on how the vector space is embedded in the algebra and what is the scalar product. As we shall see in the sequel, non-isometric vector spaces may have isomorphic Clifford algebras. In the following, we shall often speak of a Clifford algebra, but it will be understood that we consider it together with its underlying vector space and the Clifford embedding (5.5).

The subspace $K \oplus V$ of $\mathcal{T}(V)$ contains no element of $I(g)$ other than 0. Therefore, κ restricted to that subspace is injective and $K \oplus V$ may be identified with its image in $C\ell(g)$ under κ. If this is done, then, as a consequence of (5.4), one can write

$$uv + vu = 2\, g(u, v) \tag{5.6}$$

for every $u, v \in V$. Observe that two vectors anticommute if, and only if, they are orthogonal.

Theorem 5.2. ($C\ell$ is a functor). Let (V_i, g_i), $i = 1, 2$, be two vector spaces with scalar products. If $f : V_1 \to V_2$ is linear and

$$g_2(f(u), f(v)) = g_1(u, v) \text{ for every } u, v \in V,$$

then there exists a homomorphism of algebras with unity

$$C\ell(f) : C\ell(g_1) \to C\ell(g_2)$$

such that

$$C\ell(f) \circ \kappa_1 = \kappa_2 \circ f \tag{5.7}$$

where

$$\kappa_i : V_i \to C\ell(g_i)\,, i = 1, 2,$$

is the restriction to V_i of the canonical map.

This is a simple consequence of Theorem 5.1. In particular, to any orthogonal transformation $f \in O(g)$ there corresponds an automorphism $C\ell(f)$ of the algebra $C\ell(g)$ and there is the homomorphism of groups

$$C\ell : O(g) \to \mathrm{Aut}\, C\ell(g). \tag{5.8}$$

If f is the reflection, $f(v) = -v$, $v \in V$, then $\alpha = C\ell(f)$ is called the *main automorphism* of the Clifford algebra. Sometimes it is necessary to specify the vector space V underlying the Clifford algebra where the main automorphism is defined; it is then denoted by $\alpha(V)$. It is involutive and defines the basic $\mathbb{Z}_2$-grading of the Clifford algebra,

$$C\ell(g) = C\ell_0(g) \oplus C\ell_1(g)$$

where the decomposition of $a \in C\ell(g)$ into its even $a_0 \in C\ell_0(g)$ and odd $a_1 \in C\ell_1(g)$ parts is given by

$$a_0 = \frac{1}{2}(a + \alpha(a)), \qquad a_1 = \frac{1}{2}(a - \alpha(a)).$$

The subalgebra $C\ell_0(g)$ is called the *even Clifford algebra.*

Let $\mathcal{A}$ be the transposed (opposite) algebra of $C\ell(g)$: its underlying vector space is the same as that of $C\ell(g)$ and the product of a and b in $\mathcal{A}$ is equal to the product of b and a in $C\ell(g)$. The map $f : V \to \mathcal{A}$ given by $f(u) = u$ is a Clifford map and $\tilde{f} : C\ell(g) \to \mathcal{A}$ is an isomorphism of algebras with unity which may be represented as an antiisomorphism

$$\beta : C\ell(g) \to C\ell(g)$$

It is characterized as a linear isomorphism such that

$$\beta(1) = 1, \qquad \beta(u) = u \quad \text{and} \quad \beta(ab) = \beta(b)\,\beta(a), \tag{5.9}$$

for every $u \in V$ and $a, b \in C\ell(g)$. The map β will be called the *main antiautomorphism* of $C\ell(g)$. Sometimes it will be written as $\beta(V)$ to indicate the underlying vector space of the algebra.

5.2. The vector space structure of Clifford algebras

The following construction, summarized in Theorem 5.3., shows that, *as vector spaces*, the Clifford algebra of (V,g) and the Grassmann algebra ΛV are isomorphic. For every $u \in V$, let j_u be the endomorphism of ΛV given by

$$j_u(w) = u \wedge w, \quad \text{where } w \in \Lambda V .$$

If $\omega \in V^*$, then the interior product i_ω is an odd derivation (antiderivation) of ΛV characterized by

$$i_\omega(u) = <\omega, u> \quad \text{for every } u \in V$$

$$i_\omega(v \wedge w) = i_\omega(v) \wedge w + (-1)^\ell v \wedge i_\omega(w) \quad \text{for } v \in \Lambda^\ell V \text{ and } w \in \Lambda V.$$

Since

$$i_\omega \circ j_u + j_u \circ i_\omega = <\omega, u> \mathrm{id},$$

the map

$$u \to f(u) = i_{g(u)} + j_u \tag{5.10}$$

is a Clifford map from V to End ΛV. It extends, therefore, to a homomorphism

$$\tilde{f} : C\ell(g) \to \mathrm{End}\, \Lambda V$$

of algebras with unity.

Theorem 5.3. The map

$$F_g : C\ell(g) \to \Lambda V$$

given by

$$F_g(s) = <\tilde{f}(s), 1> \tag{5.11}$$

is an isomorphism of vector spaces which preserves the $\mathbb{Z}_2$-grading and is natural with respect to isometries.

Note that $\tilde{f}$, occurring in (5.11), is obtained by extending the Clifford map (5.10) according to Theorem 5.1. The right-hand side of (5.11) denotes the value of the endomorphism $\tilde{f}(s)$ of ΛV on the scalar $1 \in \Lambda V$. Therefore, in particular

$$F_g(1) = 1, \quad F_g(u) = u, \tag{5.12}$$
$$F_g(u\, v) = u \wedge v + g(u, v)$$

for every $u, v \in V$. The property of "being natural with respect to isometries" means that, if $f : V \to W$ is an isometry from (V, g) to (W, h), then $(\Lambda f) \circ F_g = F_h \circ C\ell(f)$, where $\Lambda f : \Lambda V \to \Lambda W$ is the extension of f,

$$f(1) = 1, \qquad (\Lambda f)(v_1 \wedge \ldots \wedge v_\ell) = f(v_1) \wedge \ldots \wedge f(v_\ell). \tag{5.13}$$

We refer to Bourbaki (1959) for a proof of the Theorem.

Let (e_i), $i = 1, \ldots, n$ be an orthogonal basis in (V, g), i.e. a linear basis such that

$$g(e_i, e_j) = 0 \quad \text{for } i \neq j.$$

The isomorphism F_g maps the element of $C\ell(g)$

$$e_{i_1} e_{i_2} \ldots e_{i_\ell}, \quad \text{where } 1 \leq i_1 < i_2 < \ldots < i_\ell \leq n \text{ and } 1 \leq \ell \leq n, \tag{5.14}$$

into $e_{i_1} \wedge e_{i_2} \wedge \ldots \wedge e_{i_\ell} \in \Lambda V$. Therefore, the set (5.14), supplemented with 1 (= product of an empty sequence of basis vectors), constitutes a linear basis of $C\ell(g)$ consisting of 2^n elements. The set $\{e_1, \ldots, e_n\}$ generates $C\ell(g)$ as an algebra.

The isomorphism F_g allows us to refer to the elements of the Clifford algebra by the names of their images in ΛV. Thus $uv - vu$, where $u, v \in V$, is a "bivector" and

$$\eta = e_1 e_2 \ldots e_n \tag{5.15}$$

is a "volume element".

It is often convenient to refer to elements such as $e_{i_1} \ldots e_{i_p}$ as being of degree p. It should be remembered, however, that whereas the exterior product of two elements of degrees p and q is of degree p+q, their Clifford product is, in general, a sum of elements of degrees

$$|p-q|, \quad |p-q| + 2, \ldots, p+q.$$

The main antiautomorphism β is linear and involutive,

$$\beta^2 = \mathrm{id}.$$

Therefore it defines a decomposition of $C\ell(g)$ into a direct sum of vector spaces

$$C\ell(g) = C\ell_+(g) \oplus C\ell_-(g) \tag{5.16}$$

corresponding to the decomposition

$$a = a_+ + a_- , \qquad a \in C\ell(g)$$

where

$$a_\pm = \frac{1}{2}(a \pm \beta(a))$$

so that

$$\beta(a_\pm) = \pm a_\pm .$$

If the dimension of V is larger than 1, then $C\ell_+(g)$ is not a subalgebra. It is clear from the defining properties (5.9) of β that

$$\beta(e_{i_1} \dots e_{i_p}) = (-1)^{p(p-1)/2} e_{i_1} \dots e_{i_p} . \tag{5.17}$$

The dimension of $C\ell_+(g)$ depends only on the dimension n of V. Since

$$\dim C\ell_+(g) - \dim C\ell_-(g) = \sum_{p=0}^{n} (-1)^{p(p-1)/2} \binom{n}{p}$$

we can write

$$\dim C\ell_+(g) - \dim C\ell_-(g) = 2\, d(n)$$

where

$$d(n) = 2^{(n-2)/2} \left(\cos \frac{n\pi}{4} + \sin \frac{n\pi}{4}\right) \tag{5.18}$$

is, for n ≥ 1, an integer-valued function with the property

$$d(n+8) = 16\, d(n) \tag{5.19}$$

Its first eight values are as follows

(5.20)

n	1	2	3	4	5	6	7	8
d(n)	1	1	0	-2	-4	-4	0	8

5.3 The graded structure of Clifford algebras

The importance of the grading of Clifford algebras becomes apparent when one considers their centres and the construction of a Clifford algebra of a direct sum of two vector spaces.

Consider an n-dimensional vector space V over K and its Clifford algebra $C\ell(g)$. If (e_i) is an orthogonal basis in V, then the volume element (5.15) satisfies

$$\eta u + (-1)^n u \eta = 0 \tag{5.21}$$

for every $u \in V$. Therefore, for n odd, $K \oplus K\eta$ is contained in the centre of $C\ell(g)$. More precisely, we have the following

Lemma. Let

$$Z_{\pm}(g) = \{z \in C\ell(g) \mid zu \mp uz = 0 \quad \text{for every } u \in V\}. \tag{5.22}$$

If n is even, then

$$Z_{+}(g) = K \quad \text{and} \quad Z_{-}(g) = K\eta \tag{5.23}$$

If n is odd, then

$$Z_{+}(g) = K \oplus K\eta \quad \text{and} \quad Z_{-}(g) = \{0\}. \tag{5.24}$$

To *prove* the Lemma, we first note that if $z \in C\ell(g)$ is decomposed into its even and odd parts, $z = z_0 + z_1$, then

$$z\,u \mp u\,z = 0, \qquad u \in V, \tag{5.25}$$

is equivalent to

$$z_i\, u \mp u\, z_i = 0 \quad \text{for both } i = 0 \text{ and } i = 1.$$

It suffices, therefore, to consider (5.25) for z homogeneous of degree i = 0 or 1. Let

$$z = e_1\, z' + z''$$

where z' and z'' do not contain e_1 when decomposed into the basis vectors (5.14). Putting $u = e_1$ in (5.25) we obtain

$$(1 \pm (-1)^i)\, z' = 0 \quad \text{and} \quad (1 \mp (-1)^i)\, z'' = 0 \tag{5.26}$$

Whenever the signs are such that (5.26) implies $z' = 0$ we can repeat the reasoning for $u = e_2, \dots e_n$ and infer that z is an element of K. Similarly, if (5.26) implies $z'' = 0$, then z is proportional to η. If $z \in Z_+(g) \cap Cl_1(g)$, then $z'' = 0$ and z may be any multiple of η if n is odd; $z = 0$ if n is even. If $z \in Z_+(g) \cap Cl_0(g)$, then $z' = 0$ and $z \in K$. This completes the proof the statement concerning $Z_+(g)$. If $z \in Z_-(g) \cap Cl_0(g)$ then $z'' = 0$ and z may be a multiple of η or $z = 0$ depending on whether n is even or odd. If $z \in Z_-(g) \cap Cl_1(g)$, then $z' = 0$ and $z = 0$ is the only number anticommuting with all vectors. This completes the proof of the Lemma.

It is clear that the centre of $Cl(g)$ coincides with $Z_+(g)$ and the supercentre of $Cl(g)$ is

$$SZ(Cl(g)) = Z_+(g) \cap Cl_0(g) \oplus Z_-(g) \cap Cl_1(g)$$

The Lemma leads to

Theorem 5.4. Every Clifford algebra is supercentral. The Clifford algebra of an even-dimensional space is central. The centre of a Clifford algebra of an odd-dimensional space is generated by the volume element.

Consider now the direct sum $(V \oplus W, g \oplus h)$ of two spaces with scalar products, (V, g) and (W, h). The map

$$f : V \oplus W \to Cl(g) \hat{\otimes} Cl(h) \tag{5.27}$$

given by

$$f(v, w) = v \otimes 1 + 1 \otimes w \tag{5.28}$$

is linear and has the Clifford property.

$$f(v, w)^2 = g(v, v) + h(w, w)$$

because, by the definition (4.39) of the graded tensor product, we have

$$(v \otimes 1).(1 \otimes w) + (1 \otimes w).(v \otimes 1) = 0$$

By Theorem 5.1, there is a homomorphism

$$\tilde{f} : Cl(g \oplus h) \to Cl(g) \hat{\otimes} Cl(h) \tag{5.29}$$

of algebras with unity which extends (5.27). By considering the action of f on an orthogonal basis, one infers that f is an isomorphism. This establishes the fundamental

Theorem 5.5 (Chevalley (1954). There is a natural, grading-preserving isomorphism (5.29) of algebras with unity resulting from extending the Clifford map (5.28).

If g is a scalar product in V, then -g is the *opposite* scalar product in V,

$$(-g)(u, v) = -g(u, v), \quad \text{where} \quad u, v \in V.$$

The injection

$$V \rightarrow \mathcal{C}\ell(g)^{opp}$$

is a Clifford map of (V, -g) into $\mathcal{C}\ell(g)^{opp}$. This observation leads at once to

Theorem 5.6. The algebras

$$\mathcal{C}\ell(-g) \quad \text{and} \quad \mathcal{C}\ell(g)^{opp} \tag{5.30}$$

are isomorphic as graded algebras with unity.

Assume now that g is neutral; therefore, V is even-dimensional and admits a decomposition

$$V = N \oplus P \tag{5.31}$$

into a direct sum of two totally null subspaces. Let

$$u = n + p$$

be the decomposition of a vector $u \in V$ corresponding to (5.31). If $w \in \Lambda N \subset \Lambda V$, then (cf. § 5.2)

$$i_\omega(w) \text{ and } j_n(w) \in \Lambda N$$

where

$$\omega = 2g(p) \in V^*.$$

Since both n and p are null, $g(u,u) = 2g(n,p) = \langle \omega, n \rangle$, and

$$i_\omega \circ j_n + j_n \circ i_\omega = g(u,u)\, \mathrm{id}$$

Therefore, the linear map

$$V \to \mathrm{End}\,\Lambda N$$

defined by

$$u \to i_\omega + j_n$$

has the Clifford property. It extends, therefore, to a homomorphism $\ell\colon C\ell(g) \to \mathrm{End}\,\Lambda N$ of algebras with unity. By inspecting the image by ℓ of a basis of $C\ell(g)$ generated by a null basis of V adapted to the decomposition (5.31) one proves that ℓ is an isomorphism. Finally, let End ΛN be given a grading such that $a \in \mathrm{End}\,\Lambda N$ is even (resp., odd) if it preserves (resp., changes) the parity of the degree of a homogeneous element of ΛN. Clearly, both i_ω and j_n are odd. Therefore, h preserves the grading. If V is 2m-dimensional over K, then ΛN is 2^m-dimensional over K and End ΛN is isomorphic to the matrix algebra $K(2^m)$. Its even subalgebra is isomorphic to $2K(2^{m-1})$. This can be summarized in

Theorem 5.7. Let V be a 2m-dimensional vector space over K with a neutral scalar product g. Its Clifford algebra

$$C\ell_0(g) \to C\ell(g)$$

is graded isomorphic to the matrix algebra

$$2K(2^{m-1}) \to K(2^m) \tag{5.32}$$

graded by taking the tensor product of the graded algebra

$$2\mathbb{R} \to \mathbb{R}(2), \text{ where } \sigma \text{ and } \varepsilon \text{ are odd,} \tag{5.33}$$

with the ungraded algebra $K(2^{m-1})$.

The beautiful Chevalley Theorem 5.5 is not always convenient for practical computations: the graded tensor product is less easy to use than the ordinary tensor product. For this reason, we state and prove another theorem on Clifford algebras of direct sums of vector spaces.

Let (V, g) and (W, h) be two vector spaces over K with scalar products. The tensor product of their Clifford algebras admits a natural grading such that

$$\deg(a \otimes b) \equiv \deg a + \deg b \mod 2$$

where $a \in \mathcal{C}\ell(g)$ and $b \in \mathcal{C}\ell(h)$. Assume that W is even-dimensional and let η be a volume element in that space so that $\eta w + w\eta = 0$ for every $w \in W$ and $\lambda = \eta^2$ is a number, $\lambda \neq 0$. The linear map f defined by

$$f(v, w) = v \otimes \eta \; + \; 1 \otimes \omega \tag{5.34}$$

is a Clifford map of $(V \oplus W, \lambda g \oplus h)$ into $\mathcal{C}\ell(g) \otimes \mathcal{C}\ell(h)$. By inspection of its action on a suitable basis one obtains

Theorem 5.8. Let λ denote the square of a volume element η in a 2m-dimensional space with a scalar product h. There is a graded isomorphism of algebras with unity

$$\mathcal{C}\ell(\lambda g \oplus h) \to \mathcal{C}\ell(g) \otimes \mathcal{C}\ell(h) \tag{5.35}$$

obtained by extending the Clifford map

$$f : V \oplus W \to \mathcal{C}\ell(g) \otimes \mathcal{C}\ell(h)$$

defined by (5.34). Moreover,

$$\alpha(V \oplus W) = \alpha(V) \otimes \alpha(W) \tag{5.36}$$

and

$$\beta(V \oplus W) = \begin{cases} \beta(V) \otimes \beta(W) & \\ & \text{for m} \\ \beta(V) \otimes (\alpha\,(W)\;\beta(W)) & \end{cases} \begin{matrix} \text{even} \\ \\ \text{odd} \end{matrix} \tag{5.37}$$

The last part of the Theorem follows from

$$\beta(W)\eta = (-1)^m \eta,$$

cf. (5.17). Another general result is contained in

Theorem 5.9. Let V be a real vector space with a scalar product g. There is a graded isomorphism of complex algebras with unity.

$$\mathcal{C}\ell(\mathbb{C} \otimes g) \to \mathbb{C} \otimes \mathcal{C}\ell(g)$$

obtained by extending the obvious Clifford map

$$\mathbb{C} \otimes V \to \mathbb{C} \otimes C\ell(g).$$

The proof is a straightforward application of the definition (3.38) of the complexified scalar product $\mathbb{C} \otimes g$, followed by inspection of a suitable basis.

5.4 The volume element and Hodge duality

By a suitable choice of the orthogonal basis (e_i) in V, the volume element (5.15) can be *normalized* in such a way that its square is either 1 or -1. More precisely, if V is complex, then one can always achieve $\eta^2 = 1$. If V is real of signature (k, l), then, by taking an orthonormal frame (e_i) one obtains

$$\eta^2 = (-1)^{(k-\ell)(k-\ell-1)/2} = \begin{cases} 1 & \text{if } k - \ell \equiv 0 \text{ or } 1 \bmod 4 \\ -1 & \text{if } k - \ell \equiv 2 \text{ or } 3 \bmod 4 \end{cases} \qquad (5.38)$$

The normalization condition determines η only up to a sign: choosing one of the two normalized volume elements is equivalent to introducing an orientation in the space V. From now on, all volume elements will be assumed to be normalized.

Whenever η belongs to the centre of an algebra $\mathcal{A}$ and $\eta^2 = 1$, there is a decomposition of $\mathcal{A}$ into a direct sum of subalgebras,

$$\mathcal{A} = (1 + \eta)\, \mathcal{A} \;\oplus\; (1 - \eta)\, \mathcal{A}. \qquad (5.39)$$

Therefore, the decomposition (5.39) holds, with η being a volume element

i) for the full Clifford algebra of an odd-dimensional complex vector space;
ii) for the full Clifford algebra of a real vector space of signature (k, ℓ) such that $k-\ell \equiv 1 \bmod 4$,
iii) for the even Clifford algebra of an even-dimensional complex vector space;
iv) for the even Clifford algebra of a real vector space of signature (k, ℓ) such that $k-\ell \equiv 0 \bmod 4$.

The map

$$a \mapsto \eta a$$

is an automorphism of the vector space structure of $C\ell(g)$. The element ηa is sometimes called the (Hodge) dual of a. If V is complex, then the dual of the dual, $\eta^2 a$ is equal to a. If V is real then $\eta^2 a = +a$ or $-a$, depending on the signature, cf.(5.38). The isomorphism F_g described in §5.2 can be used to define duals of elements of ΛV. It is customary in differential geometry to denote the Hodge dual by a star and to define it by

$$* F_g(a) = F_g(\eta\, \beta(a)), \quad \text{where} \quad a \in C\ell(g).$$

The dual of the dual is now

$$** F_g(a) = F_g(\eta\, a\, \beta(\eta))$$

Therefore, if V is a real space of signature (k, ℓ) and $\omega \in \Lambda^p V$, then

$$**\omega \;=\; (-1)^{(k+\ell+1)p+\ell}\; \omega \tag{5.40}$$

Let W be an m-dimensional subspace of V. Consider a basis (e_i) of V adapted to W in the sense that the vectors $e_1, ..., e_m$ form an orthogonal basis of W. The full basis (e_i) of V need not be orthogonal; in fact, if W is *null* in the sense of containing a null direction orthogonal to W, then there is no orthogonal basis in V adapted to W. Let

$$\mu = e_1\, e_2 \dots e_m$$

and consider the vector space

$$W^{\perp} \;=\; \{u \in V \mid [u, \mu] = 0\} \tag{5.41}$$

where [,] denotes the graded commutator defined by (4.36). Since

$$[u, \mu] = [u, e_1]\, e_2 \dots e_m \;-\; e_1\, [u, e_2] \dots e_m + \;\dots + (-1)^{m-1}\, e_1 e_2 \dots [u, e_m]$$

and

$$[u, v] = 2g(u, v) \quad \text{for every} \;\; u, v \in V,$$

the space $W^{\perp}$ consists of all vectors orthogonal to all elements of W. If W is null, then one can take e_1 to be in the null direction orthogonal to W. Therefore, if W is null, and only in this case,

$$\mu^2 = 0.$$

If W is not null, then the vectors $e_1, \;...,\; e_m$ spanning W can be chosen to be orthonormal and μ becomes a normalized volume element for W,

$$\mu^2 = \pm 1.$$

Let u be a vector and consider the graded commutator

$$[u, \eta\mu] = (-1)^{n-1}\, \eta\, (u\mu + (-1)^m\, \mu u)$$

where n is the dimension of V. If $u \in W$, then

$$u\, \mu + (-1)^m\, \mu\, u = 0. \tag{5.42}$$

Conversely, condition (5.42) implies $u \in W$ because, if $u \notin W$ then one can take $e_{m+1} = u$ and

$$\mu\, u + (-1)^m\, u\, \mu = 2e_1 \dots e_m\, e_{m+1} + \text{terms of lower degree}$$

is different from 0. This leads to the following characterization of the subspace W in terms of $\eta\mu$,

$$W = \{u \in V \mid [u, \eta\mu] = 0\}. \tag{5.43}$$

A subspace W is null if

$$W \cap W^{\perp} \neq \{0\}.$$

It is totally null if $W \neq \{0\}$ and

$$W \subset W^{\perp}.$$

If (V, g) is neutral, and, therefore, of even dimension 2m, then it admits totally null subspaces of (maximal) dimension m. If W is such a subspace, then

$$W = W^{\perp}$$

and the dual $\eta\mu$ is proportional to μ. Since $\eta^2 = 1$ for a neutral space, we have

$$\text{either} \quad \eta\mu = +\mu \quad \text{or} \quad \eta\mu = -\mu. \tag{5.44}$$

Depending on the sign in (5.44), the totally null subspace W is said to be of positive or negative *helicity* with respect to the orientation of V defined by η.

Example 5.1. As an illustration, consider the totally null subspaces W_0 and W_1 of K^{2m} defined in §3.5. The vectors (e_i), $i = 1, \dots, 2m$, of the null basis, considered as elements of the corresponding Clifford algebra, satisfy

$$e_\alpha \, e_{m+\beta} + e_{m+\beta} \, e_\alpha = \delta_{\alpha\beta},$$

$$e_\alpha \, e_\beta + e_\beta \, e_\alpha = 0,$$

where $\alpha, \beta = 1, \dots, m$. The vectors

$$e_1 - e_{m+1}, \dots, e_m - e_{2m}, e_1 + e_{m+1}, \dots, e_m + e_{2m}$$

constitute an orthonormal basis in K^{2m} and their product, which may be written as

$$\eta = (e_1 \, e_{m+1} - e_{m+1} \, e_1) \dots (e_m \, e_{2m} - e_{2m} \, e_m)$$

is a normalized volume element, $\eta^2 = 1$. The space W_0 (resp., W_1) is spanned by the vectors $e_{m+1}, \dots, e_{2m}$ (resp., $e_1, \dots, e_m$). Putting

$$\mu_0 = e_{2m} \dots e_{m+1} \quad \text{and} \quad \mu_1 = e_1 \dots e_m$$

we see that

$$\eta\mu_0 = (-1)^m \, \mu_0 \quad \text{and} \quad \eta\mu_1 = \mu_1.$$

The totally null subspaces W_0 and W_1 of $V = K^{2m}$ are of the same helicity when m is even and of the opposite helicity when m is odd. In particular, for $m = 2$, the pairs (e_1, e_2) and (e_3, e_4) span null 2-spaces of the same helicity which is opposite to that of the null 2-spaces spanned by the pairs (e_1, e_4) and (e_2, e_3).

Note also that the product $\nu = \mu_1\mu_0$ of the 2m vectors of a null basis in K^{2m} is not a volume element; it is an idempotent, $\nu^2 = \nu$, and the left ideal $C\ell(g)\nu$ is minimal (Chevalley 1954, Lounesto 1981).

5.5 Relation between the Clifford algebras of vector spaces of adjacent dimension

The strategy to find Clifford algebras of high dimensional spaces is to begin with low dimensions and to use theorems 5.5 and 5.8 relating the Clifford algebra of a direct sum to tensor products of the algebras. Theorem 5.8, which is of much use, is restricted by the assumption that one of the summands is even-dimensional. It is convenient to have at hand a scheme for relating to each other Clifford algebras of vector spaces differing in dimension by one. It turns out that there is an (ungraded) isomorphism between the full Clifford algebra of a vector space and the even Clifford algebra of a space of dimension larger by one; if the vector spaces are real, there is a relation between the signatures of the two spaces in question, as we now proceed to show.

Let V be a vector space over K, with a scalar product g, and let W be a one-dimensional space over the same field; its scalar product is denoted by h. Let $e_0 \in W$ be a unit vector, i.e. $\lambda = h(e_0, e_0)$ is either 1 (if $K = \mathbb{C}$ or $K = \mathbb{R}$ and h is positive) or -1 (if h is negative). The linear map

$$f : V \to C\ell_0 (g \oplus h)$$

given by

$$f(u) = e_0 u, \quad \text{where} \quad u \in V, \tag{5.45}$$

has the Clifford property,

$$f(u)^2 = -\lambda g(u, u).$$

It extends, therefore, to a homomorphism of algebras. By inspection of its action on a suitable basis, one arrives at

Theorem 5.10. Consider the spaces (V, g) and (W, h) over K, where W is one-dimensional and let $\lambda = h(e_0, e_0)$ be the square of a volume element in W; there is then an isomorphism

$$C\ell(-\lambda g) \to C\ell_0(g \oplus h) \tag{5.46}$$

of algebras with unity, obtained by extending the Clifford map f given by (5.45). The image of the even subalgebra $C\ell_0(-\lambda g)$ is generated in $C\ell_0(g \oplus h)$ by all elements of the form uv, where $u \in V$.

6. COMPLEX CLIFFORD ALGEBRAS

In this chapter we describe in detail the structure of complex Clifford algebras, i.e. of Clifford algebras associated with finite dimensional complex vector spaces with a scalar product. Considerable attention is given to the relations between representations of Clifford algebras associated with vector spaces of adjacent dimension. We exhibit three types of extensions of representations and the recurrence properties of the associated inner products. The groups of automorphisms of the inner products are also derived.

6.1 Dirac and Weyl spinors

An n-dimensional complex vector space with a scalar product is isomorphic to $(\mathbb{C}^n, g_n)$, cf. §3.5. Therefore, it is enough to find the Clifford algebras

$$\mathcal{C}\ell(n) = \mathcal{C}\ell(g_n) \qquad \text{for } n = 1, 2, \ldots$$

Since g_{2m} is neutral, it follows from Theorem 5.7 that the structure of the Clifford algebra of a complex, even-dimensional space is given by

$$\mathcal{C}\ell_o(2m) = 2\mathbb{C}(2^{m-1}) \to \mathbb{C}(2^m) = \mathcal{C}\ell(2m), \quad m = 1,2, \ldots \tag{6.1}$$

According to Theorem 5.10,

$$\mathcal{C}\ell(n) = \mathcal{C}\ell_o(n+1) \tag{6.2}$$

and the structure of the Clifford algebra of a complex, (2m+1)-dimensional space is given by the double of $\mathbb{C}(2^m)$,

$$\mathcal{C}\ell_o(2m+1) = \mathbb{C}(2^m) \to 2\,\mathbb{C}(2^m) = \mathcal{C}\ell(2m+1), \quad m = 0, 1, \ldots \ . \tag{6.3}$$

The algebras $\mathcal{C}\ell(2m)$ and $\mathcal{C}\ell_0(2m+1)$ are simple. Each of them admits, up to equivalence, only one faithful irreducible representation (Theorem 4.1). The elements of the complex carrier space of such a representation are called *Dirac spinors*. According to this definition, the space of Dirac spinors associated with both $(\mathbb{C}^{2m}, g_{2m})$ and $(\mathbb{C}^{2m+1}, g_{2m+1})$ is complex 2^m-dimensional. This terminology seems to have been accepted in the physics literature, cf., for example, Coquereaux (1985) and the references given there. Note, however, that — as defined here — the Dirac spinors for $(\mathbb{C}^3, g_3)$ are complex 2-component: they are what one often calls Pauli spinors. Since there are important differences in the structure of the Clifford algebras of even- and odd-dimensional spaces, one could make a case for using a different name (e.g. Pauli's), for spinors associated with odd dimensions.

Let

$$\gamma : \mathcal{C}\ell(2m) \rightarrow \mathrm{End}\, S \tag{6.4}$$

be a faithful irreducible representation of $\mathcal{C}\ell(2m)$ in the space S of Dirac spinors. Let η be the volume element and

$$\Gamma = \gamma(\eta)$$

the corresponding endomorphism of S. Since η is normalized, $\eta^2 = 1$, we have

$$\Gamma^2 = I$$

and there is the direct sum decomposition

$$S = S_+ \oplus S_- \tag{6.5}$$

where

$$S_\pm = \{\phi \in S \mid \Gamma\phi = \pm\,\phi\}.$$

Since η is in the centre of $\mathcal{C}\ell_0(2m)$, the restriction γ_0 of γ to $\mathcal{C}\ell_0(2m)$ decomposes,

$$\gamma_0 = \gamma_+ \oplus \gamma_-\,,$$

where

$$\gamma_\pm(a) = \frac{1}{2}(I \pm \Gamma)\,\gamma(a) \qquad \text{for every } a \in \mathcal{C}\ell_0\,(2m).$$

The algebra $\mathcal{C}\ell_0(2m)$ has two representations

$$\gamma_\pm : \mathcal{C}\ell_0(2m) \rightarrow \mathrm{End}\, S_\pm \tag{6.6}$$

which are 2^{m-1}-dimensional, irreducible and inequivalent, but not faithful. The elements of S_+ and S_- are called *Weyl spinors* of positive and negative *helicities*, respectively. Physicists often refer to them as right and left Weyl spinors. Cartan (1938) and Bourbaki (1959) use the expression semi spineurs. Chevalley (1954) calls them half-spinors. Penrose and Rindler (1986) describe S_+ and S_- as the reduced spin spaces.

Note that when the orientation is changed, i.e. when η is replaced by $-\eta$, the roles of right and left spinors are interchanged.

Given the representation (6.4) of $C\ell(2m)$, one can construct a faithful representation γ' of $C\ell(2m+1)$ as follows. Let e_{2m+1} be a vector in $\mathbb{C}^{2m+1} = \mathbb{C}^{2m} \oplus \mathbb{C}$ orthogonal to $\mathbb{C}^{2m}$ and such that $e^2_{2m+1} = 1$. We define two irreducible representations γ'_+ and γ'_- of $C\ell(2m+1)$ in S by

$$\gamma'_\pm(v) = \pm\gamma(v) \quad \text{and} \quad \gamma'_\pm(e_{2m+1}) = \pm\Gamma, \tag{6.7}$$

where $v \in \mathbb{C}^{2m}$. The volume element associated with $\mathbb{C}^{2m+1}$ is

$$\eta' = \eta \; e_{2m+1}$$

and, since $\gamma'_\pm(\eta) = \Gamma$, we have

$$\gamma'_\pm(\eta') = \pm I$$

so that none of the representations γ'_+ and γ'_- is faithful. Their direct sum,

$$\gamma' = \gamma'_+ \oplus \gamma'_- : C\ell(2m+1) \to \text{End S} \oplus \text{End S}, \tag{6.8}$$

is a faithful representation of $C\ell(2m+1)$ in $S \oplus S$. The representations γ'_+ and γ'_- are equal when restricted to $C\ell_0(2m+1)$. These restrictions provide a faithful irreducible representation which, with a slight abuse of notation, we denote also by γ. This representation

$$\gamma : C\ell_0(2m+1) \to \text{End S} \tag{6.9}$$

coincides with (6.4) on $C\ell_0(2m)$ and is given by

$$\gamma(e_{2m+1}\, a) = \Gamma\, \gamma(a) \quad \text{for } a \in C\ell_1(2m). \tag{6.10}$$

6.2 The inner products

Since the complex algebra $C\ell(2m)$ is simple and endowed with the main antiautomorphism β, one can apply to it the results of § 4.3 to construct an inner product B on the space of Dirac spinors. It follows from (4.14) that

$${}^t(B\,\gamma(a)) = ({}^tB\, B^{-1})\, B\, \gamma(\beta(a))$$

for every a $\in$ $C\ell(2m)$. Therefore, if $\beta(a) = a$, i.e. a $\in$ $C\ell_+(2m)$, then the symmetry of B γ (a) is the same as that of B. Since γ is faithful and B non-degenerate, the map a $\rightarrow$ B γ (a) is a linear isomorphism of $C\ell(2m)$ on the vector space

product in the space of Weyl spinors and

$$ {}^tB_+ = \begin{cases} B_- & \text{for } m \equiv 1 \mod 4 \\ -B_- & \text{for } m \equiv 3 \mod 4 \end{cases} \tag{6.15} $$

For every Clifford algebra $C\ell(g)$ of a vector space V over K, in addition to the map B intertwining the representations γ and $\check{\gamma}$, one can also consider the map

$$ E : S \to S^* $$

such that

$$ {}^t\gamma\,(\beta(a))\,E = E\,\gamma(\alpha(a)) \tag{6.16} $$

where $a \in C\ell(g)$ and α is the main automorphism; equivalently,

$$ {}^t\gamma\,(u) = -E\,\gamma\,(u)\,E^{-1} $$

for every $u \in V$. If V is even-dimensional and $\Gamma = \gamma(\eta)$, then[1)]

$$ E = B\Gamma. \tag{6.17} $$

In particular, for the algebra $C\ell(2m)$, it follows from (6.11) and (6.13) that

$$ {}^tE = \begin{cases} E & \text{for } m = 0,3 \mod 4, \\ -E & \text{for } m = 1,2 \mod 4. \end{cases} \tag{6.18} $$

Since the algebra $C\ell(2m+1)$ is not simple, one cannot directly apply to it the general results of § 4.3. For *even* m, however, it follows from (6.7) and (6.13) that

$$ {}^t\gamma_\pm\,(\beta(a))\,B = B\,\gamma_\pm(a) \quad \text{for } a \in C\ell(2m+1), \tag{6.19} $$

and, therefore, B provides an inner product on the space of Dirac spinors. If m is *odd*, then the representation γ_+ is equivalent to γ_-,

---

1) Brauer and Weyl (1935) and Cartan (1938) used the linear map $C : S \to S^*$ such that ${}^t\gamma\,(u) = (-1)^m\,C\,\gamma\,(u)\,C^{-1}$, where $\gamma : C\ell(2m)$ or $C\ell_0(2m+1) \to \mathrm{End}\ S$ and u is a vector. Therefore, their C coincides with our $B\Gamma^m = E\Gamma^{m+1}$. We prefer to reserve the letter C for a semi-linear map $S \to S$ associated with charge conjugation.

$$ {}^t\gamma_+ (\beta(a))\, B\Gamma = B\Gamma\, \gamma_-(a) \qquad \text{for } a \in \mathcal{C}\ell(2m+1), \tag{6.20} $$

therefore

$$ {}^t\gamma\, (\beta(a))\, B\Gamma = B\Gamma\, \gamma\,(a) \qquad \text{for } a \in \mathcal{C}\ell_o(2m+1), $$

and we have an inner product BΓ on the carrier space of the representation (6.9). It can be extended, "antidiagonally", to the carrier space $S \oplus S$ of the representation (6.8) of $\mathcal{C}\ell(2m+1)$.

6.3 Extensions of representations

There are several "natural" extensions to $\mathcal{C}\ell(2m+2)$ of the representations of $\mathcal{C}\ell(2m)$ and $\mathcal{C}\ell(2m+1)$ described above. Even though all faithful irreducible representations of a simple algebra are equivalent, for specific applications one may prefer one representation or another. These extensions allow a simple, inductive construction of the (generalized) Dirac matrices for higher-dimensional spaces and of the inner products on the spaces of Dirac spinors. Before giving a description of the extensions, we recall a few facts about the algebra $\mathbb{C}(2)$. It contains two matrices N and P such that

$$ NP + PN = 0 \quad \text{and} \quad P^2 = -N^2 = I. $$

The matrix $M = NP$ anticommutes with both N and P, and $M^2 = I$. Any matrix $X \in \mathbb{C}(2)$ is a linear combination of I, M, N and P; moreover, the matrices M, N, P are traceless and

$$ \operatorname{Tr} X = 0 \quad \text{iff} \quad {}^tX = -\varepsilon X \varepsilon^{-1}, \tag{6.21} $$

where ε is as in (4.2).

Given a representation (6.4), we construct its extension

$$ \gamma' : \mathcal{C}\ell(2m+2) \to \operatorname{End}(S \oplus S) \tag{6.22} $$

by requiring, firstly, that there should be a matrix $X \in \mathbb{C}(2)$ such that

$$ \gamma'(a) = X \otimes \gamma(a) \tag{6.23} $$

for every $a \in \mathcal{C}\ell(2m)$; we have identified here the algebras $\operatorname{End}(S \oplus S)$ and $\mathbb{C}(2) \otimes \operatorname{End} S$. Consider now the subalgebra $\mathcal{A}$ of $\mathcal{C}\ell(2m+2)$ generated by the vector space $\mathbb{C}^2$, which is being added to $\mathbb{C}^{2m}$, and the subalgebra $\mathcal{B}$ of End $(S \oplus S)$ generated by all elements of the form $N \otimes I$, $P \otimes I$ and $I \otimes \Gamma \in \mathbb{C}(2) \otimes \text{End } S$. The algebras $\mathcal{A}$ and $\mathcal{B}$ are isomorphic to $\mathbb{C}(2)$ and $2\mathbb{C}(2)$, respectively. The extension is completed by specifying a suitable homomorphism of $\mathcal{A}$ into $\mathcal{B}$. Such *simple extensions* may be classified according to the dimension of the intersection of $\gamma'(\mathbb{C}^2)$ with the vector subspace of $\mathcal{B}$ spanned by all elements of the form $Y \otimes \Gamma$, where $Y \in \mathbb{C}(2)$. There are three types of extensions, corresponding to the dimensions 0, 1 and 2. They will be called here extensions of *type* 0, 1 and 2, respectively.

To describe explicitly the extensions it is convenient to use an orthogonal basis (e_α) in $\mathbb{C}^n$ such that

$$e_\alpha^{\,2} = (-1)^{\alpha+1}, \qquad \alpha = 1, \ldots, n,$$

so that the fundamental quadratic form is

$$u_1^2 - u_2^2 + \ldots - (-1)^n u_n^2 .$$

For every of the representations γ we introduce the "Dirac matrices"

$$\gamma_\alpha = \gamma(e_\alpha)$$

Their product

$$\Gamma = \gamma_1 \cdots \gamma_n$$

satisfies

$$\Gamma^2 = I.$$

Consider first an extension of type 0. Since, in this case, $\gamma'(u) = U \otimes I$ for $u = x\, e_{2m+1} + y e_{2m+2}$, where $U \in \mathbb{C}(2)$ anticommutes with X of (6.23), we are led to the form of the extension presented in the first column of Table I. From ${}^t(P \otimes \gamma_\alpha) = -\varepsilon P \varepsilon^{-1} \otimes B \gamma_\alpha B^{-1}$, ${}^t(M \otimes I) = -\varepsilon M \varepsilon^{-1} \otimes I$ and ${}^t(N \otimes I) = -\varepsilon N \varepsilon^{-1} \otimes I$, we obtain that

$$B' = \varepsilon P \otimes B \quad \text{intertwines } \gamma' \text{ and } \check{\gamma}' \text{ of type 0.}$$

Similarly, for type 1, we may choose e_{2m+1} and e_{2m+2} so that $\gamma'(e_{2m+1}) = M \otimes \Gamma$ and $\gamma'(e_{2m+2}) = N \otimes I$; the matrix X must then commute with M and anticommute with N; this leads to the Dirac matrices presented in the second column. By virtue of (6.13), the intertwining isomorphisms B' and E' depend now on the parity of m. The extension of type 2 is obtained in a similar manner.

In all three cases, there is still freedom to choose the anticommuting matrices M, N and P. Of special interest is *type 1* which we discuss in some detail:

(i) The choice $M = \tau$, $N = \varepsilon$ made by *Brauer and Weyl* (1935) has the property that γ' restricted to $C\ell(2m+1)$ coincides with the representation (6.8).

(ii) The *Cartan extension* is obtained by taking $M = \sigma$ and $N = \varepsilon$ so that $\Gamma' = \tau \otimes I$ and the two subspaces occurring in the direct sum $S \oplus S$ coincide with the two spaces of Weyl spinors associated with $C\ell_0(2m+2)$.

(iii) Another possibility is to take $M = i\,\varepsilon$ and $N = i\,\tau$; we call it the *Dirac extension* because, if it is applied to the representation $\gamma : C\ell(2) \to \mathbb{C}(2)$ with $\gamma_1 = \sigma\ (=\sigma_x)$ and $\gamma_2 = -i\,\varepsilon\ (=\sigma_y)$, it leads, up to factors of i, to the standard representation of the gamma matrices used in low-energy quantum mechanics of the electron (Dirac (1928).

For each of the three types, there is a choice of the matrices M and N such that the representation γ' restricted to $C\ell(2m)$ splits into a direct sum of representations in the two subspaces S of $S \oplus S$. (One chooses $P = \tau$ for type 0, $M = \tau$ for type 1; in type 2 this holds for every choice of the matrices). This being so, one can say that the Dirac spinor of $C\ell(2m+2)$ is the sum of two Dirac spinors of $C\ell(2m)$; such a splitting is convenient for the discussion of spinors in higher-dimensional extensions of space-time. On the other hand, essentially only the Cartan extension allows the interpretation of elements of S as Weyl spinors; for this reason it is convenient for considerations of high-energy physics.

Let γ' and γ'' be any two representations chosen among the extended representations of type 0, 1 and 2. There is then a simple isomorphism

$$F : \mathbb{C}^2 \otimes S \to \mathbb{C}^2 \otimes S \tag{6.24}$$

intertwining γ' and γ'',

$$\gamma''(a)\,F = F\,\gamma'(a).$$

Indeed, the isomorphism

$$F = \tfrac{1}{2}\, I \otimes (I + \Gamma) + \tfrac{1}{2}\, N \otimes (I - \Gamma) \tag{6.25a}$$

intertwines the representations of type 0 and 1, and the isomorphism

$$F' = \tfrac{1}{2}\, I \otimes (I + \Gamma) + \tfrac{1}{2}\, M \otimes (I - \Gamma) \tag{6.25b}$$

intertwines the representations of type 1 and 2. Within each type, there are intertwining isomorphisms (6.24) of a rather special form, with $F = G \otimes I$ and $G \in GL(2, \mathbb{C})$. For example if γ' is the Brauer-Weyl extension, then $G = I - \varepsilon$ and $I - i\sigma$ connect it to the Cartan and Dirac extensions γ'', respectively.

Table I

Simple extensions of a representation γ: $C\ell(2m) \to \text{End } S$ to representations γ': $C\ell(2m+2) \to \mathbb{C}(2) \otimes \text{End } S$

$\alpha = 1, \ldots, 2m$	Type 0	Type 1	Type 2
γ'_α γ'_{2m+1} γ'_{2m+2}	$P \otimes \gamma_\alpha$ $M \otimes I$ $N \otimes I$	$M \otimes \gamma_\alpha$ $M \otimes \Gamma$ $N \otimes I$	$I \otimes \gamma_\alpha$ $M \otimes \Gamma$ $N \otimes \Gamma$
Γ'	$P \otimes \Gamma$	$P \otimes I$	$P \otimes \Gamma$
B'	$\varepsilon P \otimes E$	$\varepsilon P \otimes B$ m even $\varepsilon M \otimes E$ m odd	$\varepsilon P \otimes B$ m even $\varepsilon \otimes B$ m odd
E'	$\varepsilon \otimes B$	$\varepsilon \otimes B$ m even $\varepsilon N \otimes E$ m odd	$\varepsilon \otimes E$ m even $\varepsilon P \otimes E$ m odd

Notation: $M, N, P \in \mathbb{C}(2)$, $NP + PN = 0$, $P^2 = I = -N^2$, $M = NP$,
$^t\gamma_\alpha = B\,\gamma_\alpha\,B^{-1} = -E\,\gamma_\alpha\,E^{-1}$, $\Gamma = \gamma_1 \ldots \gamma_{2m}$, and similarly for the primed quantities.

The convenience of the Cartan extension justifies giving an explicit form of the gamma matrices and inner products which can be obtained by induction from

$$\gamma'_\alpha = \sigma \otimes \gamma_\alpha, \quad \gamma'_{2m+1} = \sigma \otimes \Gamma, \quad \gamma'_{2m+2} = \varepsilon \otimes I.$$

For m = 1 we can choose

$$\gamma_1 = \sigma \quad \text{and} \quad \gamma_2 = \varepsilon$$

and then obtain the explicit form of the gamma matrices in the *Cartan representation* for the vector space $\mathbb{C}^{2m}$:

$$\gamma_{2\alpha-1} = \sigma \otimes \ldots \otimes \sigma \otimes \tau \otimes I \otimes \ldots \otimes I$$
$$\gamma_{2\alpha} = \sigma \otimes \ldots \otimes \sigma \otimes \varepsilon \otimes I \otimes \ldots \otimes I \tag{6.26}$$

where $\alpha = 1, \ldots, m$ and each of the tensor products contains m-α matrices σ and α-1 unit 2 by 2 matrices I. The corresponding product of the Dirac matrices is

$$\Gamma = \tau \otimes I \otimes \ldots \otimes I \quad \text{(m-1 unit matrices)}.$$

Since, in this case,

$${}^t\gamma_\alpha = (-1)^{\alpha+1}\, \gamma_\alpha,$$

we can take

$$B = \gamma_2\, \gamma_4 \cdots \gamma_{2m}\, \Gamma^m,$$
$$E = \gamma_1\, \gamma_3 \cdots \gamma_{2m-1}\, \Gamma^m.$$

Explicitly,

$$B = \begin{cases} \Sigma_m & \text{for} \quad \text{m even,} \\ \sigma \otimes \Sigma_{m-1} & \text{for} \quad \text{m odd,} \end{cases} \tag{6.27}$$

where

$$\Sigma_{2p} = (\tau \otimes \varepsilon) \otimes \ldots \otimes (\tau \otimes \varepsilon)$$

is the tensor product of p factors, each equal to $\tau \otimes \varepsilon$.

The symmetries of B, given by (6.11), exhibit a periodicity of period 8 in the dimension $n = 2m$ of the underlying vector space. It is convenient to collect in a table the symmetry properties of B and E, as well as their properties related to helicity. This is presented in Table II in a notation influenced by the usage of computer programming: starting in any dimension (line of the programme), one constructs the matrices B and E in subsequent dimensions, according to the rules given in the Table, remembering that, as soon as an equality such as $B = F(E)$ appears, the B's in the following lines should be replaced by F(E), until a new formula for B is reached, which then becomes operative, etc. One can start the programme at any line, by inserting the appropriate B and E. For example, the choice $B = \sigma$ and $E = \varepsilon$ in line 3 leads to (6.27).

Note that for n odd, there is a preference to use B for $m \equiv 0, 2 \bmod 4$ and E for $m \equiv 1, 3 \bmod 4$: these inner products commute with Γ and, as a result of this are defined within each of the two spaces of Dirac spinors. The other inner products are represented by "antidiagonal " matrices.

6.4 The Wall groups

In § 4.3 we defined the Wall group $\mathcal{G}$ of the pair $(\mathcal{A}, \beta)$ where β is an involutive antiautomorphism of the algebra $\mathcal{A}$. For a Clifford algebra, we may take β to be the main antiautomorphism and then $\mathcal{G}$ is the group of automorphisms of the inner product B. But we may also replace β by $\alpha \circ \beta$ and then $\mathcal{G}$ becomes the group of automorphisms of E, cf. (6.16). For simplicity, we consider here, in every dimension, only one of the two groups which can be so defined; a complete list of the automorphism groups of B and E has been given by Lounesto (1981).

We define the Wall group $\mathcal{G}(n)$ to be the set of all elements a of the complex Clifford algebra $\mathcal{C}\ell(n)$ such that

$$\beta(a)\, a = 1 \qquad \text{for } m \text{ even}$$

and

$$\alpha \circ \beta(a)\, a = 1 \qquad \text{for } m \text{ odd,}$$

Table II

Inner products on the spaces of Dirac spinors associated with the Cartan representations of the Clifford algebras $C\ell(n)$, n = 2m or 2m+1

line	n mod 8	m mod 4	B		E	
1	0	0	$B = \begin{pmatrix} E & O \\ O & -E \end{pmatrix}$	sym.	$\begin{pmatrix} E & O \\ O & E \end{pmatrix}$	sym.
2	1	0	$\begin{pmatrix} B & O \\ O & B \end{pmatrix}$	sym.	$\begin{pmatrix} O & B \\ B & O \end{pmatrix}$	sym.
3	2	1	$\begin{pmatrix} O & B \\ B & O \end{pmatrix}$	sym.	$E = \begin{pmatrix} O & -B \\ B & O \end{pmatrix}$	skew
4	3	1	$\begin{pmatrix} O & E \\ E & O \end{pmatrix}$	skew	$\begin{pmatrix} E & O \\ O & E \end{pmatrix}$	skew
5	4	2	$B = \begin{pmatrix} E & O \\ O & -E \end{pmatrix}$	skew	$\begin{pmatrix} E & O \\ O & E \end{pmatrix}$	skew
6	5	2	$\begin{pmatrix} B & O \\ O & B \end{pmatrix}$	skew	$\begin{pmatrix} O & B \\ B & O \end{pmatrix}$	skew
7	6	3	$\begin{pmatrix} O & B \\ B & O \end{pmatrix}$	skew	$E = \begin{pmatrix} O & -B \\ B & O \end{pmatrix}$	sym.
8	7	3	$\begin{pmatrix} O & E \\ E & O \end{pmatrix}$	sym.	$\begin{pmatrix} E & O \\ O & E \end{pmatrix}$	sym.
GO TO LINE 1						

where n = 2m or 2m+1. By reference to Table II it can be seen that this particular definition leads to all the groups $\mathcal{G}(2m+1)$ being direct products of two isomorphic factors operating, via the representation (6.8), in the two spaces of Dirac spinors associated with $C\ell(2m+1)$. Writing 2G instead of G x G, denoting by p an integer divisible by 4 and $q = 2^p$, we can derive from Table II the following list of Wall groups of complex Clifford algebras:

Table III

The Wall groups of complex Clifford algebras

n	$\mathcal{G}(n)$
2p	$O(q, \mathbb{C})$
2p+1	$2\ O(q, \mathbb{C})$
2p+2	$Sp(2q, \mathbb{C})$
2p+3	$2\ Sp(2q, \mathbb{C})$
2p+4	$Sp(4q, \mathbb{C})$
2p+5	$2\ Sp(4q, \mathbb{C})$
2p+6	$O(8q, \mathbb{C})$
2p+7	$2\ O(8q, \mathbb{C})$
$p = 0 \bmod 4,\ q = 2^p$	

If m is even, then the representation γ restricted to $\mathcal{G}(n)$ coincides with the representation *contragredient* to the restriction of γ to $\mathcal{G}(n)$,

$$\check{\gamma}(a) = {}^t\gamma(a^{-1}), \quad a \in \mathcal{G}(n).$$

For m odd, $\check{\gamma}$ corresponds to a twisted contragredient representation with respect to γ.

7. REAL CLIFFORD ALGEBRAS

7.1 The index periodicity

The Clifford algebras of real vector spaces can be easily determined by finding the algebras of low-dimensional spaces and then applying an inductive procedure based on Theorems 5.5-10.

Let $C\ell(k,\ell)$ denote the Clifford algebra of the vector space $\mathbb{R}^{k+\ell}$ with the scalar product $g_{k,\ell}$ (§3.5). The vectors e_α ($\alpha = 1, ..., k+\ell$) of an orthonormal basis for $g_{k,\ell}$ satisfy

$$e_\alpha e_\beta + e_\beta e_\alpha = 0 \quad \text{for } \alpha \neq \beta,$$

$e_\alpha{}^2 = 1$ for $\alpha = 1, ..., k$ and $e_\alpha{}^2 = -1$ for $\alpha = k+1, ..., k+\ell$. Sometimes a different labelling of the vectors will be used.

If η is the volume element, $\eta = e_1 ... e_{k+\ell}$, then its square,

$$\eta^2 = (-1)^{(k-\ell)(k-\ell-1)/2} \tag{7.1}$$

depends only on the index $k-\ell$ of the scalar product.

A statement such as "the algebras $C\ell_0(k,\ell) \to C\ell(k,\ell)$ and $\mathcal{A}_0 \to \mathcal{A}$ are isomorphic as graded algebras with unity" will be shortened to "$C\ell(k,\ell)$ is isomorphic to $\mathcal{A}_0 \to \mathcal{A}$" and sometimes even to "$C\ell(k,\ell)$ and $\mathcal{A}$ are isomorphic".

Since the scalar product opposite to $g_{k,\ell}$ is $g_{\ell,k}$, from Theorem 5.6 applied to real Clifford algebras one obtains

$$C\ell(k,\ell) \quad \text{and} \quad C\ell(\ell,k)^{\text{opp}} \quad \text{are isomorphic.} \tag{7.2}$$

We first recognize that the graded algebras described in Examples 4.6-8 are isomorphic to some of the real Clifford algebras:

(i) The algebra $C\ell(0,1)$ is isomorphic to the algebra $\mathbb{R} \to \mathbb{C}$ described in Example 4.6. An isomorphism is obtained by sending e_1 to $\sqrt{-1}$.

(ii) Example 4.9 together with (7.1) provides an isomorphism between $C\ell(1,0)$ and $\mathbb{R} \to 2\mathbb{R}$.

(iii) The algebra $C\ell(2,0)$ is isomorphic to $\mathbb{C} \to \mathbb{R}(2)$ described in Example 4.7a. An isomorphism is obtained by extending $e_1 \mapsto \sigma$ and $e_2 \mapsto \tau$.

(iv) Similarly, the algebra $C\ell(1,1)$ is isomorphic to $2\mathbb{R} \to \mathbb{R}(2)$ described in Example 4.7b. This result can be also obtained from Theorem 5.7 (the scalar product $g_{1,1}$ is neutral).

(v) The algebra $C\ell(0,2)$ is isomorphic to $\mathbb{C} \to \mathrm{H}$. This may be checked either directly from $e_1 \mapsto j$, $e_2 \mapsto k$, or by appealing to Example 4.8 and Theorem 5.5. Example 4.11 may now be interpreted as providing the isomorphism between $C\ell(0,2)$ and $C\ell(2,0)^{opp}$.

Let us now apply Theorem 5.10 to real Clifford algebras by taking $g = g_{k,\ell}$. If $h = g_{1,0}$ then $\lambda = 1$ and one obtains the (ungraded) isomorphism

$$C\ell(\ell, k) = C\ell_0(k+1, \ell) \tag{7.3}$$

The other possibility, $h = g_{0,1}$ leads to $\lambda = -1$ and

$$C\ell(k, \ell) = C\ell_0(k, \ell+1). \tag{7.4}$$

By comparing (7.2) and (7.3) one obtains the isomorphism of the algebras,

$$C\ell_0(k, \ell) = C\ell_0(\ell, k)$$

even though the algebras $C\ell(k, \ell)$ and $C\ell(\ell, k)$ are not, in general, isomorphic. There is a graded version of the isomorphism (7.4):

Theorem 7.1 If k and ℓ are integers such that $k \geq 0$ and $\ell \geq 1$, then there is an isomorphism f of graded algebras

$$\begin{array}{ccc} C\ell_0(k, \ell) & \rightarrow & C\ell(k, \ell) \\ \downarrow & & \downarrow f \\ C\ell(k, \ell-1) & \underset{h}{\rightarrow} & C\ell_0(k, \ell+1) \end{array} \tag{7.5}$$

obtained by extending the Clifford map

$$f : \mathbb{R}^{k+\ell} \to C\ell_0(k, \ell+1)$$

where

$$f(u) = u\, e_{k+\ell+1} \quad \text{and} \quad u = \Sigma\, u_\alpha e_\alpha .$$

The injection (monomorphism of algebras) h is an extension of the Clifford map

$$h : \mathbb{R}^{k+\ell-1} \to C\ell_0(k, \ell+1)$$

where

$$h(u) = u\, e_{k+\ell} \quad \text{and} \quad u = \Sigma\, u_\alpha e_\alpha .$$

It is understood that $C\ell(0,0) = \mathbb{R}$ and $\mathbb{R}^0 = \{0\}$.

The grading of the algebra $C\ell_0(k, \ell+1)$ implied by (7.5) is such that an element of the algebra is odd if, and only if, it contains $e_{k+\ell+1}$ as a factor when decomposed into a sum of products of the basis vectors. The isomorphism (7.3) leads to an analogous theorem covering the case $k \geq 1$ and $\ell \geq 0$. One obtains it by replacing in the lower left-hand corner of (7.5) the algebra $C\ell(k, \ell-1)$ by $C\ell(\ell, k-1)$ and suitably modifying f and h.

According to Theorem 5.7, the neutral algebra

$$C\ell_0(m, m) \to C\ell(m, m)$$

is isomorphic to

$$2\mathbb{R}(2^{m-1}) \to \mathbb{R}(2^m). \tag{7.6}$$

Since, in this case, $\eta^2 = 1$, Theorem 5.8 applied to $g = g_{k,\ell}$ and $h = g_{m,m}$ gives the isomorphism

$$C\ell(k+m,\ \ell+m) = C\ell(k,\ell) \otimes \mathbb{R}(2^m). \tag{7.7}$$

As at least one of the two numbers k and ℓ is ≥ 1, we can use (7.3) or (7.4) to obtain

$$C\ell_0(k+m, \ell+m) = C\ell_0(k,\ell) \otimes \mathbb{R}(2^m). \tag{7.8a}$$

Let us now apply Theorem 5.8 to $g = g_{k,\ell}$ and $h = g_{2,0}$ or $g_{0,2}$. In both cases $\lambda = -1$. By referring to the examples (iii) and (v) described above, we obtain

$$C\ell(\ell+2, k) = C\ell(k,\ell) \otimes \mathbb{R}(2) \tag{7.8b}$$

and

$$C\ell(\ell, k+2) = C\ell(k,\ell) \otimes H. \tag{7.9}$$

By a repeated application of (7.7) for m = 1, of (7.8b) and (7.9) one arrives at the isomorphism

$$C\ell(k+4,\ \ell) \ = \ C\ell(k, \ell + 4) \tag{7.10}$$

and similarly for the even subalgebras. Computing $C\ell(k+4,\ \ell+4)$, using (7.7) and (7.10) one arrives at the periodicity property of real Clifford algebras,

$$C\ell(k+8,\ \ell) \ = \ C\ell(k,\ \ell+8) \ = \ C\ell(k,\ell) \otimes \mathbb{R}(16) \tag{7.11}$$

and similarly for the even subalgebras.

We can now extend the list (i)-(v) to include a few more algebras:

(vi) From (7.8) we obtain $C\ell(3,0) = C\ell(0,1) \otimes \mathbb{R}(2) = \mathbb{C} \otimes \mathbb{R}(2) = \mathbb{C}(2)$ and (7.3) leads to $C\ell_0(3,0) = C\ell(0,2) = H$.

(vii) Similarly, from (7.9) we have $C\ell(0,3) = C\ell(1,0) \otimes H = 2H$ and $C\ell_0(0,3) = C\ell_0(3,0) = H$. Therefore $C\ell(5,0) = C\ell(1,4) = C\ell(0,3) \otimes \mathbb{R}(2) = 2H(2)$.

(viii) From (7.3), (7.8) and (7.10) we obtain $C\ell(4,0) = C\ell(0,4) = H(2)$ and $C\ell_0(4,0) = C\ell_0(0,4) =$ $=2H$.

(ix) Finally, from (7.10) and (i)-(vii) we obtain the isomorphism

$$\begin{array}{ll} C\ell(5,0) = 2H(2), & C\ell(0,5) = \mathbb{C}(4). \\ C\ell(6,0) = H(4), & C\ell(0,6) = \mathbb{R}(8). \\ C\ell(7,0) = \mathbb{C}(8), & C\ell(0,7) = 2\mathbb{R}(8). \end{array}$$

Consider now the following list of eight graded real algebras[1)]

---

1) This list may be found, in a rather different notation, in E. Cartan's (1908) article in the Encyclopédie des Sciences Mathématiques.

Table IV

ν	0	1	2	3	4	5	6	7
$\mathcal{A}_\nu$	$\mathbb{R}(2)$	$2\mathbb{R}$	$\mathbb{R}(2)$	$\mathbb{C}(2)$	$\mathrm{H}(2)$	$2\mathrm{H}$	H	$\mathbb{C}$
$\mathcal{A}_{\nu o}$	$2\mathbb{R}$	$\mathbb{R}$	$\mathbb{C}$	H	$2\mathrm{H}$	H	$\mathbb{C}$	$\mathbb{R}$
n_ν	2	1	2	3	4	3	2	1

where n_ν is such that the (real) dimension of $\mathcal{A}_\nu$ is 2^{n_ν} and the gradings are as described in Ch.4. The list is arranged in such a way that there are graded isomorphisms

$$C\ell(1,1) = \mathcal{A}_0,$$

$$C\ell(\nu, 0) = \mathcal{A}_\nu \otimes \mathbb{R}\,(2^{\frac{1}{2}(\nu - n_\nu)}) \tag{7.12}$$

and

$$C\ell(0,8-\nu) = \mathcal{A}_\nu \otimes \mathbb{R}\,(2^{4-\frac{1}{2}(\nu + n_\nu)}) \tag{7.13}$$

where $\nu = 1, \ldots, 7$. They follow from (i)-(ix).

Theorem 7.2 Let k and ℓ be non-negative integers such that $n = k+\ell > 0$. There is an isomorphism f of graded algebras

$$\begin{array}{ccc} C\ell_0(k,\ell) & \rightarrow & C\ell(k,\ell) \\ \downarrow & & \downarrow f \\ \mathcal{A}_{\nu o} \otimes \mathbb{R}(2^p) & \rightarrow & \mathcal{A}_\nu \otimes \mathbb{R}\,(2^p) \end{array} \tag{7.14}$$

where the integers ν and p are given by

$$k - \ell = 8q + \nu, \quad 0 \le \nu \le 7, \quad q \in \mathbb{Z},$$
$$k + \ell = 2p + n_\nu,$$

and n_ν, $\mathcal{A}_\nu$ and $\mathcal{A}_{\nu o}$ are as in Table IV.

The proof is a straightforward application of previous results: if $\nu = 0$ and $m \geq 0$ then $p = \ell+4m-1$ and

$$C\ell(k,\ell) = C\ell(\ell+8m, \ell) = \mathbb{R}(2^{\ell+4m}) = \mathcal{A}_0 \otimes \mathbb{R}(2^p)$$

as asserted; the case $m < 0$ is treated similarly. If $\nu \neq 0$ and $k > \ell$, then

$$C\ell(k,\ell) = C\ell(\nu, o) \otimes \mathbb{R}(2^{\ell+4m}) = \mathcal{A}_\nu \otimes \mathbb{R}(2^p)$$

by virtue of (7.12). If $k < \ell$, then one uses (7.13).

Referring to Theorem 5.5 one obtains

$$\mathcal{A}_\mu \hat{\otimes} \mathcal{A}_\nu = \mathcal{A}_\rho \otimes \mathbb{R}(2^{\frac{1}{2}(n_\mu + n_\nu - n_\rho)}) \tag{7.15}$$

where ρ is the reduction of $\mu+\nu$ mod 8, i.e. $0 \leq \rho \leq 7$ and either $\mu + \nu = \rho$ or $\mu + \nu = 8 + \rho$.

Consider now the set of all graded algebras of the form $\mathcal{A}_\nu \otimes \mathbb{R}(2^p)$ where $0 \leq \nu \leq 7$ and p is an integer. They form eight classes corresponding to the eight values of ν. The graded tensor product induces a multiplication in the set of these eight classes which makes it into a group, the *Brauer-Wall group* of the reals (Wall 1964, Lounesto 1981). This group is isomorphic to $\mathbb{Z}_8$ and the class of $\mathcal{A}_0$ is its neutral element. If $[\mathcal{A}]$ denotes the class of the algebra $\mathcal{A}$ so that

$$[\mathcal{A}_\mu] \cdot [\mathcal{A}_\nu] = [\mathcal{A}_\mu \hat{\otimes} \mathcal{A}_\nu]$$

and

$$[C\ell(k,k)] = [\mathcal{A}_0] = 1$$

then

$$[C\ell(k,\ell)] = [C\ell(\ell,k)]^{-1}$$

because $g_{k,\ell} \oplus g_{\ell,k}$ is neutral. The formula for the graded tensor product (4.40) may be now interpreted as $\mathcal{A}_7 \hat{\otimes} \mathcal{A}_7 = \mathcal{A}_6$. Conversely, from (7.15) one can read off the graded tensor products of all algebras occurring in Table IV. For example, taking the square of $\mathcal{A}_6$, one gets

$$H \hat{\otimes} H = H(2) \tag{7.16}$$

This should be contrasted with the ungraded product formula

$$H \otimes H = \mathbb{R}(4) \tag{7.17}$$

established in §4.1.

Two real Clifford algebras belong to the same class,

$$[C\ell(k,\ell)] = [C\ell(k', \ell')],$$

if, and only if, their indices are congruent mod 8,

$$k-\ell \equiv k' - \ell' \mod 8.$$

Note also that in our notation formula (7.1) can be written as

$$\eta^2 = (-1)^{\nu(\nu-1)/2} \tag{7.1'}$$

7.2 Charge conjugation and Majorana spinors

Essentially all relevant information about the real Clifford algebras and their representations is contained in Table IV. It is useful, however, to present this information in a different form, adapted to the needs of theoretical physics, where complex representations are of primary significance.

Let us first note that, since the complexification of the real algebra $C\ell(k,\ell)$ is isomorphic to the complex algebra $C\ell(k+\ell)$, there is an injection

$$C\ell(k,\ell) \to C\ell(k+\ell)$$

which can be used to construct a complex representation of $C\ell(k,\ell)$ from a representation of $C\ell(k+\ell)$. If the latter representation is faithful and irreducible, then so is its restriction to $C\ell(k,\ell)$. Assume now that $k+\ell = 2m$ is *even* and let

$$\gamma : C\ell(k,\ell) \to \mathrm{End}_{\mathbb{C}}\, S \tag{7.18}$$

be a representation obtained in this manner from one of the representations described in Chapter 6. The vector space S is therefore, of complex dimension 2^m. The algebra $C\ell(k,\ell)$ is isomorphic to a matrix algebra $L(N)$ with a suitable N and $L = \mathbb{R}$ or H. It is, therefore, central simple over $\mathbb{R}$ and, by Theorem 4.2, there exists a $\mathbb{C}$-linear isomorphism $C : S \to \bar{S}$ intertwining the representations γ and $\bar{\gamma}$,

$$\bar{\gamma}(a)\, C = C\, \gamma(a) \quad \text{for every } a \in C\ell(k,\ell). \tag{7.19}$$

and such that

$$\text{either } \bar{C}C = I \quad \text{or} \quad \bar{C}C = -I$$

In the first case the representation γ is real and in the second it is quaternionic. By inspection of Table IV we obtain

$$\bar{C}C = \begin{cases} I & \text{for} \quad \nu = 0, 2 \quad \text{(real case)}, \\ -I & \text{for} \quad \nu = 4, 6 \quad \text{(quaternionic case)}. \end{cases}$$

In the *real* case, the *real form of γ decomposes*,

$$\gamma = \gamma^+ \oplus \gamma^-,$$

where

$$\gamma^{\pm} : C\ell(k,\ell) \to \mathrm{End}_{\mathbb{R}}\, S^{\pm}, \quad k - \ell \equiv 0, 2 \bmod 8,$$

are two real equivalent representations in the real vector spaces S^+ and S^-, defined as in §3.4 and §4.4. In the *quaternionic* case, the *real form of γ is irreducible* and its commutant is isomorphic to H.

The restriction γ_0 of (7.18) to the even algebra $C\ell_0(k,\ell)$ decomposes

$$\gamma_0 = \gamma_+ \oplus \gamma_-$$

where γ_+ and γ_- are the representations in the spaces S_+ and S_- of Weyl spinors defined as the two eigenspaces of the *helicity endomorphism*

$$\Gamma = i^{\nu(\nu-1)/2}\, \gamma(\eta) \tag{7.20}$$

defined so that $\Gamma^2 = I$. The complex conjugate space $\bar{S}$ also decomposes,

$$\bar{S} = \bar{S}_+ \oplus \bar{S}_-$$

where

$$\bar{S}_{\pm} = \{\bar{\phi} \in \bar{S} \mid \bar{\Gamma}\, \bar{\phi} = \pm\, \bar{\phi}\}.$$

Since (7.19) and (7.20) imply

$$\bar{\Gamma} C = (-1)^{\nu(\nu-1)/2} C \Gamma \qquad (7.21)$$

we see that C preserves or changes helicity depending on whether $\nu = 0, 4$ or $2, 6$, respectively. In other words, there is a complex equivalence of representations of $C\ell_0(k,\ell)$

$$\gamma_\pm \sim \bar{\gamma}_\pm \qquad \text{for } \nu = 0, 4,$$

and

$$\gamma_\pm \sim \bar{\gamma}_\mp \qquad \text{for } \nu = 2, 6.$$

The representations γ_+ and γ_- are never complex-equivalent to each other because

$$\gamma_\pm\ (\eta) = \pm\, i^{-\nu(\nu-1)/2}\, I. \qquad (7.22)$$

However, for $\nu = 2$ and 6 their *real forms are equivalent* (cf. §4.4): in this case the algebra $C\ell_0(k,\ell)$ is a simple — though not central — real algebra and, as such, has only one, up to equivalence, irreducible faithful real representation.

For $k + \ell$ *odd*, Theorem 7.1 can be used to reduce the problem of determining the properties of representations of $C\ell(k,\ell)$ to that for even dimensions. If $k + \ell = 2m + 1$, then the even algebra $C\ell_0(k,\ell)$ is central simple and its faithful irreducible representation γ in the complex 2^m-dimensional space S is real for $\nu = 1, 7$ and quaternionic for $\nu = 3, 5$. The full algebra $C\ell(k,\ell)$ is a direct sum of two simple algebras for $\nu = 1, 5$; it is simple — but not central — for $\nu = 3, 7$. The representation

$$\gamma : C\ell_0(k,\ell) \to \text{End } S$$

can be extended to two representations of $C\ell(k,\ell)$ in S, also denoted by γ_+ and γ_-, by setting

$$\gamma_\pm\ (\eta) = \pm\, i^{\nu(\nu-1)/2}\, I, \qquad (7.23a)$$

where η is now the normalized volume element in $C\ell(k,\ell)$, $k + \ell = 2m + 1$. Note that, with our notational conventions, we have the three representations γ, γ_+ and γ_- for every signature (k, ℓ). Their meaning, however, is rather different in even- and odd-dimensional spaces.

An alternative and equivalent way of constructing the representations of $C\ell(k,\ell)$ for $k + \ell = 2m+1$ is to start with a representation

$$\gamma' : C\ell(k,\ell+1) \to \text{End } (S \oplus S)$$

of the simple algebra $C\ell(k,\ell+1)$ in a 2^{m+1}-dimensional space and to consider the decomposition

$$\gamma'_0 = \gamma'_+ \oplus \gamma'_- : C\ell_0(k,\ell+1) \rightarrow \text{End } S \oplus \text{End } S$$

of the restriction γ'_0 of γ' to the even subalgebra.

Let $f : C\ell(k,\ell) \rightarrow C\ell_0(k,\ell+1)$ be the isomorphism defined in Theorem 7.1. The composed maps

$$\gamma'_\pm \circ f : C\ell(k,\ell) \rightarrow \text{End } S$$

are irreducible representations of the algebra such that, by virtue of (7.22),

$$\gamma'_\pm \circ f(\eta) = \pm\, i^{-\nu'(\nu'-1)/2}\, I \qquad (7.23b)$$

where now $\nu' = \nu - 1 = k - \ell - 1 \mod 8$. Since $\nu = 1 \mod 2$ and $\nu' - \nu = 1 \mod 8$, we have $\nu(\nu-1) + \nu'(\nu' - 1) = 0 \mod 8$; therefore the righthand sides of (7.23a) and (7.23b) are equal and there is an equivalence of representations, $\gamma_\pm \sim \gamma'_\pm \circ f$.

Let

$$\gamma_0 = \gamma_+ \oplus \gamma_- : \; C\ell(k,\ell) \rightarrow \text{End } S \oplus \text{End } S$$

be the direct sum of representations so obtained. Since $C\ell_0(k,\ell)$ is central simple, the representations γ and γ are equivalent,

$$\bar{\gamma}(a)\, C = C\, \gamma(a) \quad \text{for} \quad a \in C\ell_0(k,\ell)$$

and, by virtue of (7.23a) there is a complex equivalence of representations

$$\gamma_\pm \sim \bar{\gamma}_\pm \quad \text{for} \quad \nu = 1, 5$$

and (7.24a)

$$\gamma_\pm \sim \bar{\gamma}_\mp \quad \text{for} \quad \nu = 3, 7.$$

Therefore, there is also a complex equivalence $\gamma_0 \sim \gamma_0$,

$$\bar{\gamma}_0(a)\, C_0 = C_0\, \gamma_0(a) \quad \text{for} \quad a \in C\ell_0(k,\ell), \qquad (7.24b)$$

where

$$C_0 = \begin{cases} I \otimes C & \text{for } \nu = 1, 5, \\ \sigma \otimes C & \text{for } \nu = 3, 7. \end{cases} \tag{7.24c}$$

The representations γ_+ and γ_- are inequivalent for $\nu = 1$ and 5. In these cases γ_0 provides a faithful representation of the full algebra. For $\nu = 3$ and 7 the representations γ_+ and γ_- are equivalent over $\mathbb{R}$, but not over $\mathbb{C}$. They both provide faithful irreducible representations of the full algebra in the 2^m-dimensional complex vector space S.

To establish a relation between C and the notion of *charge conjugation* of a wave function, we consider the generalized Dirac equation in a space-time $\mathbb{R}^{2m}$ with a flat metric g of signature (k,ℓ) with $k+\ell = 2m$, $m = 1, 2, \ldots$. Let (e_α) be the vectors of an orthonormal basis, $g_{\alpha\beta} = g(e_\alpha, e_\beta)$ and $\gamma_\alpha = \gamma(e_\alpha)$, where γ is a representation (7.18) of $C\ell(k,\ell)$, in S. We put $\gamma^\alpha = g^{\alpha\beta}\gamma_\beta$, where $(g^{\alpha\beta})$ is the inverse of $(g_{\alpha\beta})$, and consider a particle of mass κ and charge e moving in an electromagnetic field derived from a potential A_α. The wave function of the particle

$$\psi : \mathbb{R}^{2m} \to S$$

is assumed to be a solution of the *Dirac equation* which can be written as either

$$(\gamma^\alpha (\partial_\alpha - ieA_\alpha) - \kappa)\, \psi = 0 \tag{7.25}$$

or

$$(i\gamma^\alpha (\partial_\alpha - ieA_\alpha) - \kappa)\, \psi = 0. \tag{7.25i}$$

For a free particle ($A_\alpha = 0$) one derives the Klein-Gordon equations:

$$(g^{\alpha\beta}\, \partial_\alpha \partial_\beta - \kappa^2)\, \psi = 0 \quad \text{from (7.25)}$$

and

$$(g^{\alpha\beta}\, \partial_\alpha \partial_\beta + \kappa^2)\, \psi = 0 \quad \text{from (7.25i).}$$

The choice between (7.25) and (7.25i) depends on the signature and involves a decision concerning the signs in the equation $g^{\alpha\beta}\, p_\alpha\, p_\beta = \pm\, \kappa^2$: which of them corresponds to real particles of momentum p, rather than tachyons. In particular, in Minkowski space-time $\mathbb{R}^4$ one chooses (7.25) or (7.25i) depending on whether the signature is (3,1) or (1,3).

By complex conjugation, using

$$\bar{\gamma}_\alpha = C\, \gamma_\alpha\, C^{-1} \quad \text{and} \quad \Gamma\, \gamma_\alpha + \gamma_\alpha\, \Gamma = 0,$$

one derives

$$(\gamma_\alpha (\partial_\alpha + ie A_\alpha) - \kappa) C^{-1} \bar{\psi} = 0 \quad \text{from (7.25)}$$

and

$$(i\gamma^\alpha (\partial_\alpha + ie A_\alpha) - \kappa) (C\Gamma)^{-1} \bar{\psi} = 0 \quad \text{from (7.25i).}$$

Therefore, if ψ is a solution of (7.25), then the charge conjugate wave function

$$\psi_c = \vec{C} \bar{\psi} \tag{7.26}$$

is a solution of the same equation with opposite charge. Similarly, if ψ is a solution of (7.25i), then the charge conjugate wave function is

$$\psi_{ci} = \bar{C}_i \bar{\psi}, \quad \text{where } C_i = C\Gamma. \tag{7.26i}$$

When $\bar{C}C = I$, i.e. for $\nu = 0$ and 2, there is the decomposition (3.17), where the spaces S^+ and S^- may be now characterized by

$$S^{\pm} = \{\phi \in S \mid \phi_c = \pm \phi\}. \tag{7.27}$$

The spaces S^+ and S^- are each real 2^m-dimensional and the representation γ of the full Clifford algebra is real (§4.4): when restricted to S^+ or S^-, and expressed with respect to a basis, it is given by *real* 2^m by 2^m matrices. Moreover, *for* $\nu = 2$, the spaces S^+ and S^- have a natural complex structure given by the endomorphism $\gamma(\eta)$. The $\mathbb{C}$-linear isomorphisms

$$F_{\pm} : S_{\pm} \to S^{\pm}, \tag{7.28}$$

where

$$F_{\pm}(\phi) = \phi \pm \phi_c \quad \text{and } F_{\pm}^{-1}(\psi) = \tfrac{1}{2} (I \pm \Gamma) \psi \tag{7.29}$$

intertwine the representations $\gamma_{\pm}$ and $\gamma^{\pm}$ restricted to $C\ell_0(k,\ell)$. Since in this case $\Gamma\bar{C} = -C\Gamma$, there are no common eigenvectors of C and Γ and, therefore, no real Weyl spinors, $S_{\pm} \cap S^{\pm} = \{0\}$ for every combination of the signs.

When $\bar{C}\bar{\Gamma} C \Gamma = I$, i.e. for $\nu = 0$ and 6, there is another decomposition of the space of Dirac spinors,

$$S = S_i^+ \oplus S_i^-,$$

where

$$S_i^{\pm} = \{\phi \in S \mid \phi_{ci} = \pm \phi\} \tag{7.27i}$$

are real spaces which, however, are not invariant with respect to the representation γ: since γ_α and T anticommute, the endomorphisms γ_α map S_i^+ into S_i^-. Moreover, if $\phi \in S_i^-$ then $i\phi \in S_i^+$. Therefore, the endomorphisms $i\gamma_\alpha$ map S_i^+ into S_i^+ and are represented by real matrices with respect to a basis in S_i^+. The endomorphisms γ_α are thus represented by *pure imaginary* 2^m by 2^m matrices. The replacement of γ_α by ${}_i\gamma_\alpha = i\gamma_\alpha$ is equivalent to going from a representation γ of $C\ell(k,\ell)$ in S to a representation

$$ {}_i\gamma : C\ell(\ell,k) \rightarrow \text{End } S \tag{7.30} $$

of the opposite algebra $C\ell(\ell,k)$ and it is worth recalling in this context that the classes $[\mathcal{A}_2]$ and $[\mathcal{A}_6]$ are opposite elements of the Brauer-Wall group. The elements of S^+ and S_i^+ are sometimes called *Majorana spinors* of the *first* and *second kind*, respectively (Regge 1984). We follow this terminology even though the remarks made above show that the Majorana spinors of the second kind, associated with $C\ell(k,\ell)$, $k-\ell = 6 \bmod 8$, are equivalent to those of the first kind associated with $C\ell(\ell,k)$. For this reason, it is legitimate to refer, as is done in most of the literature, only and simply to *Majorana spinors*.

By comparing equations (7.25 and 25i) with their complex conjugate, one sees that Majorana wave functions correspond to particles without electric charge.

For $\nu = 6$ there is a complex structure in both S_i^+ and S_i^- given by $\gamma(\eta)$. The $\mathbb{C}$-linear isomorphisms

$$ F_{\pm i} : S_\pm \rightarrow S_i^\pm \tag{7.28i} $$

where

$$ F_{\pm i}(\phi) = \phi \pm \phi_c \quad \text{and} \quad F_{\pm i}^{-1}(\psi) = \tfrac{1}{2}\,(I \pm C\Gamma)\,\psi \tag{7.29i} $$

intertwine $\gamma_\pm$ and the reductions $\gamma_i^\pm$ of $\gamma \mid C\ell_0(k,\ell)$ to $S_i^\pm$.

The intertwining isomorphisms (7.28) and (7.28i) provide an equivalence of the Weyl to the Majorana spinors of the first and second kind, respectively (Pauli 1957, Gürsey 1958). We should also note the equivalence of γ^+ and γ^- and of the associated spaces of Majorana spinors of both kinds. Let $K : S^+ \rightarrow S^-$ be the "multiplication by i" map intertwining γ^+ and γ^- (§4.4), then

$$ F_- \circ L = K \circ F_+, $$

where

$$ L : S_+ \rightarrow S_-, \quad L(\phi) = i\phi_c, $$

is an isomorphism intertwining the representations γ_+ and γ_- of the even algebra.

The algebras with $\nu = 2$ and 6 are of special interest in physics because they correspond to the Minkowski signatures (3,1) and (1,3), respectively, as well as to their extensions (k+2, k) and (k, k+2), k = 2,3, ..., appearing in conformal geometry and twistor theory (Penrose 1967, Budinich 1979).

For $\nu = 0$ all three representations γ, γ_+ and γ_- are real and there are Weyl-Majorana spinors of both kinds. Defining

$$S^+ \cap S_i^+ = S_+^+ , \quad S^+ \cap S_i^- = S_-^+$$

$$S^- \cap S_i^- = S_+^- , \quad S^- \cap S_i^+ = S_-^- \ .$$

we obtain a decomposition of the real form of S into four real 2^{m-1}-dimensional spaces:

$$S = S_+^+ \oplus S_-^+ \oplus S_+^- \oplus S_-^- \ .$$

Since $\eta^2 = 1$, there is no natural complex structure in this case.

For $\nu = 4$ there are no Majorana spinors of any kind and the two Weyl representations γ_+ and γ_- are inequivalent. All three representations γ, γ_+ and γ_- are quaternionic.

Example 7.1 Let $\gamma : C\ell(1,1) \rightarrow \text{End } S$, where $S = \mathbb{C}^2$, be the representation (cf. Example 4.7b) defined by

$$\gamma_1 = \sigma, \quad \gamma_2 = \varepsilon \quad \text{so that} \quad \Gamma = \tau \quad \text{and } C = I.$$

The space

$$S^+ = \{\phi \in S \mid \bar{\phi} = \phi\}$$

is simply $\mathbb{R}^2 \subset \mathbb{C}^2$. Majorana spinors of the first kind are elements of $\mathbb{C}^2$ with both components real. On the other hand, the space of Majorana spinors of the second kind,

$$S_i^+ = \{\phi \in S \mid \tau\bar{\phi} = \phi\}$$

consists of all pairs $(\lambda, i\mu)$, where $\lambda, \mu \in \mathbb{R}$. As a linear basis in S_i^+ one can take the vectors ϕ_1 and $i\phi_2$, where

$$\phi_1 = \begin{pmatrix} 1 \\ 0 \end{pmatrix} \text{ and } \phi_2 = \begin{pmatrix} 0 \\ 1 \end{pmatrix}$$

The endomorphisms γ_1 and γ_2 are represented, with respect to the new basis, by pure imaginary matrices $i\varepsilon$ and $i\sigma$, respectively.

Example 7.2 (Clifford algebras of Minkowski space).
(i) Consider the algebra $C\ell(3,1)$ represented in $S = \mathbb{C}^4$ by

$$\gamma_1 = \tau \otimes \sigma, \quad \gamma_2 = \tau \otimes \tau, \quad \gamma_3 = \sigma \otimes I, \quad \gamma_4 = \varepsilon \otimes I$$

Here $\nu = 2$, $C = I$ and Majorana spinors (of the first kind) are real elements of $\mathbb{C}^4$. However, $\Gamma =$ $= i\tau \otimes \varepsilon$ is complex and so are Weyl spinors:

$$S_\pm = \{\lambda\, \phi_1 \otimes \phi_\mp + \mu\, \phi_2 \otimes \phi_\pm \mid \lambda, \mu \in \mathbb{C}\},$$

where $\phi_\pm = \phi_1 \pm i\phi_2$, and ϕ_1, ϕ_2 are the canonical basis vectors of $\mathbb{C}^2$ defined in the preceding Example. We can also make explicit the action of $F_+ : S_+ \to S^+ = \mathbb{R}^4 \subset \mathbb{C}^4 = S$,

$$F_+(\lambda\phi_1 \otimes \phi_\mp + \mu\phi_2 \otimes \phi_\pm) = \mathrm{Re}\,\lambda\, \phi_1 \otimes \phi_1 + \mathrm{Im}\,\lambda\, \phi_1 \otimes \phi_2 + \mathrm{Re}\,\mu\, \phi_2 \otimes \phi_1 - \mathrm{Im}\,\mu\, \phi_2 \otimes \phi_2.$$

(ii) The algebra $C\ell(1,3)$ can be represented in $S = \mathbb{C}^4$ by the Dirac matrices

$$\gamma_0 = \tau \otimes I, \quad \gamma_1 = \varepsilon \otimes \sigma, \quad \gamma_2 = -i\varepsilon \otimes \varepsilon, \quad \gamma_3 = \varepsilon \otimes \tau.$$

Here $\nu = 6$, $C = -\tau \otimes \varepsilon$, $\Gamma = \sigma \otimes I$, $C\Gamma = \varepsilon \otimes \varepsilon$ and there are Majorana spinors of the second kind,

$$S_i^+ = \{\lambda\, \phi_+ \otimes \phi_+ - \bar{\lambda}\, \phi_- \otimes \phi_- + \mu\, \phi_+ \otimes \phi_- + \bar{\mu}\, \phi_- \otimes \phi_+ \mid \lambda, \mu \in \mathbb{C}\}.$$

The map F_{+i} transforms a Weyl spinor $(\phi_1 + \phi_2) \otimes \psi$, where $\psi \in \mathbb{C}^2$, into a Majorana spinor of the second kind.

Table V

Number and dimensions of irreducible complex and real representations of $C\ell(k,\ell)$ and $C\ell_0(k,\ell)$

0	ν		0	1	2	3	4	5	6	7
1	$C\ell$	ℂ	1	2	1	2	1	2	1	2
2		ℝ	1	2	1	1+	1+	2+	1+	1+
3	$C\ell_0$	ℂ	2-	1	2-	1	2-	1	2-	1
4		ℝ	2-	1	1	1+	2	1+	1	1

Explanation: $k+\ell = 2m$ or $2m+1$ and $k-\ell = 8q + \nu$, where $0 \leq \nu \leq 7$. The figure 1 or 2 appearing in rows from 1 to 4 is the number of complex-inequivalent (rows 1 and 3) or real-inequivalent (rows 2 and 4) representations of $C\ell(k,\ell)$ and $C\ell_0(k,\ell)$. If there is no sign after that figure, then the representations in question are each of dimension 2^m over ℂ (rows 1 and 3) or ℝ (rows 2 and 4). The signs + and - indicate that those dimensions are 2^{m+1} and 2^{m-1}, respectively.

7.3 The Dirac forms

Clifford algebras of real, *even* -dimensional vector spaces satisfy the assumptions of Theorem 4.2 and their faithful irreducible representations are equivalent to their Hermitean conjugates. If $k+\ell = 2m$, then the faithful irreducible representation (7.18) is equivalent to the *Hermitean conjugate* representation

$$\gamma^\dagger : C\ell(k,\ell) \to \mathrm{End}_{\mathbb{C}}\, \overline{S}^*, \quad \text{where } \gamma^\dagger(a) = \gamma(\beta(a))^\dagger \quad \text{and} \quad a \in C\ell(k,\ell). \tag{7.31a}$$

According to Theorem 4.2,

$$\gamma(\beta(a))^\dagger A = A\, \gamma(a), \tag{7.31b}$$

where the isomorphism

$$A = \overline{B}C \tag{7.32}$$

can be made Hermitean, $A = A^\dagger$, by a choice of B and C. If, in the definition of $\gamma^\dagger$, the antiautomorphism β is replaced by $\alpha \circ \beta$, then one obtains another representation equivalent to γ,

$$\gamma(\alpha \circ \beta(a))^\dagger D = D\; \gamma(a), \tag{7.33}$$

where

$$D = i^\ell\, \overline{E}\, C \tag{7.34}$$

is also Hermitean by virtue of (4.32), (6.13), (7.21) and the congruence

$$\ell \equiv m + \tfrac{1}{2}\, \nu\, (\nu - 1) \bmod 2 \tag{7.35}$$

valid for all integers k and ℓ, with $k+\ell = 2m$ or $2m+1$.

Let ${}_i\gamma$ be the representation (7.30) of the opposite algebra $C\ell(\ell,k)$. Denoting with a subscript i the intertwining isomorphisms associated with ${}_i\gamma$, we have

$$\Gamma_i = (-1)^m \Gamma, \quad A_i = D, \quad B_i = i^{-\ell} B, \quad C_i = C\Gamma, \quad D_i = A, \quad E_i = i^\ell E. \tag{7.36}$$

where the phase factor in B_i is chosen so that $B_i C_i$ be Hermitean.

The isomorphisms A and D are used to define the *Dirac Hermitean forms* on the space S of spinors,

$$A(\phi, \psi) = \langle \bar{A}\,(\bar{\phi}), \psi\rangle \quad \text{and} \quad D(\phi, \psi) = \langle \bar{D}\,(\bar{\phi}), \psi\rangle, \tag{7.37}$$

where $\phi, \psi \in S$. By virtue of (7.32) and (7.34) the Dirac forms can also be written as

$$A(\phi, \psi) = B\,(\phi_c, \psi) \quad \text{and} \quad D(\phi, \psi) = i^{-\ell} B\,(\phi_{ci}, \psi). \tag{7.38}$$

These forms have been introduced by Dirac (1928) in connection with the equation of the electron. To establish the relation between our notation and the one in current usage in the physics literature, let us consider, for example, the Dirac form D and introduce a basis (e_μ) in the spinor space S. In agreement with the conventions of spinor calculus (§3.3), the corresponding bases in S^*, $\bar{S}$ and $\bar{S}^*$ are (e^μ), $(e^{\dot\mu})$ and $(e_{\dot\mu})$, respectively, where $\mu = 1, ..., 2^m$. Since D is Hermitean, the matrix $(D_{\mu\nu})$ of its components with respect to such a basis

$$D_{\dot\mu\nu} = D\,(e_\mu, e_\nu),$$

is also Hermitean,

$$D_{\dot\mu\nu} = D_{\dot\nu\mu}.$$

The value of D on the pair (ϕ, ψ) of spinors is

$$D(\phi, \psi) = \phi^{\dot\mu}\, D_{\dot\mu\nu}\, \psi^\nu, \quad \text{where } \phi^{\dot\mu} = \overline{\phi^\mu}. \tag{7.39}$$

The spinor

$$\bar{D}(\bar{\phi}) = \phi^{\dot\mu}\, D_{\dot\mu\nu}\, e^\nu \in S^*$$

is often called the *Dirac adjoint* of ϕ with respect to D and denoted by $\bar{\phi}$; we cannot follow this notational tradition because bar has been reserved here for complex conjugation; instead, we use a tilde, and write

$$\tilde{\phi} = \bar{D}(\bar{\phi}) \quad \text{or} \quad \tilde{\phi}_\nu = \phi^{\dot\mu}\, D_{\dot\mu\nu}\,.$$

Since D is Hermitean, $\bar{D} = {}^tD$, and (7.39) can be written as

$$\langle\tilde{\phi}, \psi\rangle = \langle\bar{\phi}, D\psi\rangle$$

or simply

$$\tilde{\phi}\,\psi = \phi^\dagger D\psi$$

in a notation which is as close as it can be to that of the majority of physicists under the provision that, for us, bar denotes complex conjugation.

In signature (3,1), the Lagrangian for (7.25) is proportional to

$$\mathrm{Re}\, D(\psi, \mathrm{Dir}\, \psi), \tag{7.40}$$

where $\psi : \mathbb{R}^4 \to \mathbb{C}^4$ is the wave function and

$$\mathrm{Dir}\, \psi = (\gamma^\alpha (\partial_\alpha - ieA_\alpha) - \kappa)\, \psi$$

is the left hand side of the Dirac equation. The correctness of (7.40) is based on $(D\gamma_\alpha)^\dagger = -D\gamma_\alpha$, an equation which leads to

$$D(\phi, \mathrm{Dir}\, \psi) - D(\mathrm{Dir}\, \phi, \psi) = \partial_\alpha D(\phi, \gamma^\alpha \psi), \tag{7.41}$$

i.e. to the formal self-adjointness of Dir, so that the first variation of the real function (7.40) is

$$\mathrm{Re}\, (2\, D(\delta\psi, \mathrm{Dir}\psi) + \partial_\alpha D(\psi, \gamma^\alpha \delta\psi)).$$

Similarly, in signature (1,3), the Lagrangian for (7.25i) is

$$\mathrm{Re}\, A\, (\psi, \mathrm{Dir}_i\, \psi) \tag{7.42}$$

where $\mathrm{Dir}_i\psi$ is the left hand side of (7.25i). If ${}_i\gamma$ is the representation used in that equation, then A in (7.42) is identified with $A_i = D$ and the Lagrangian (7.42) becomes identical to (7.40).

The following properties of the matrices A, D and γ_α are of frequent use: let $1 \le \alpha_1 < \alpha_2 \ldots < \alpha_p \le 2m$, then

$$(A\, \gamma_{\alpha_1} \ldots \gamma_{\alpha_p})^\dagger = (-1)^{p(p-1)/2}\, A\gamma_{\alpha_1} \ldots \gamma_{\alpha_p} \tag{7.43}$$

and

$$(D\, \gamma_{\alpha_1} \ldots \gamma_{\alpha_p})^\dagger = (-1)^{p(p+1)/2}\, D\gamma_{\alpha_1} \ldots \gamma_{\alpha_p} \tag{7.44}$$

What happens to the Dirac forms on restriction to the spaces of Weyl and Majorana spinors, if any? From (6.13), (7.21) and (7.35) one derives

$$A\Gamma = (-1)^k \, \Gamma^\dagger A \tag{7.45}$$

and a similar equation for D. Let $\phi = \phi_+ + \phi_-$ be the decomposition of $\phi \in S$ into Weyl spinors $\Gamma\phi_\pm = \pm\phi_\pm$, then

$$A(\phi, \psi_+) = \begin{cases} A\,(\phi_+, \psi_+) & \text{for } k \text{ even,} \\ A\,(\phi_-, \psi_+) & \text{for } k \text{ odd,} \end{cases} \tag{7.46}$$

and similarly for D. Therefore, *for even k, the Dirac forms restrict to Hermitean and non-singular forms on the spaces of Weyl spinors. For odd k,* these restrictions are zero: *the maps A and D change helicity* of Weyl spinors; such is the case of Minkowski space.

The hermicity of A and D, together with (7.21) and

$$^tB = (-1)^{m(m-1)/2}\, B, \tag{7.47}$$

implies

$$\bar{A}C = (-1)^{m(m-1)/2}\, {}^tCA \tag{7.48}$$

and

$$\bar{D}C = (-1)^{m(m+1)/2}\, {}^tCD. \tag{7.49}$$

Let $\nu = 0$ or 2, so that $\bar{C}C = I$ and there is the decomposition (3.17) of S into spaces of Majorana spinors of the first kind S^+ and S^-. From (7.38) or (7.48) we obtain that the form

$$i^{m(m-1)/2}\, A \text{ restricted to } S^\pm \text{ is real.} \tag{7.50}$$

This bilinear form is proportional to $B|S^\pm$ and, therefore, is symmetric for m = 0,1 mod 4 and skew for m = 2,3 mod 4. There are similar statements concerning D and Majorana spinors of the second kind.

The definition of Dirac forms on spinor spaces associated with real *odd*-dimensional vector spaces requires special attention because the algebra $C\ell(k,\ell)$ is not central simple for $k+\ell = 2m+1$. Let γ_+ and γ_- be the representations of the full algebra defined by the extension (7.23a) of the representation γ of the even subalgebra. By virtue of (6.19) and (6.20), there is the equivalence of representations,

$$\check{\gamma}_{\pm} \sim \gamma_{\pm} \quad \text{for m even,}$$

$$\check{\gamma}_{\pm} \sim \gamma_{\mp} \quad \text{for m odd.}$$

By comparing this with (7.24a) and using the congruence (7.35) we recognize that for ℓ even there exists a Hermitean map A such that

$$\gamma_{\pm}(\beta(a))^{\dagger} A = A\, \gamma_{\pm}(a), \quad a \in C\ell(k,\ell). \tag{7.51}$$

Similarly, for ℓ odd there is a Hermitean D such that

$$\gamma_{\pm}(\alpha \circ \beta(a))^{\dagger} D = D\, \gamma_{\pm}(a), \quad a \in C\ell(k,\ell). \tag{7.52}$$

7.4 Clifford algebras of Euclidean spaces

A real vector space is said to be Euclidean if it is given a definite scalar product.

Theorem 7.3. Let $\mathcal{A}$ be either $C\ell(k,\ell)$ or $C\ell_0(k,\ell)$ depending on whether $k+\ell = 2m$ or $2m+1$ and let γ be a faithful irreducible representation of $\mathcal{A}$ in a complex space S of dimension 2^m. The scalar products h and h' on $\mathcal{A}$,

$$h(a, b) = 2^{-m}\, \mathrm{Tr}\, \gamma(\beta(a)b) \tag{7.53}$$

$$h'(a, b) = 2^{-m}\, \mathrm{Tr}\, \gamma(\alpha \circ \beta(a)b), \quad \text{where } a, b \in \mathcal{A}, \tag{7.54}$$

are real and

h is positive-definite for $k > 0$ and $\ell = 0$; it is neutral otherwise;

h' is positive-definite for $k = 0$ and $\ell > 0$; it is neutral otherwise.

Proof. In view of the isomorphisms (7.3 and 4) it is enough to consider the case of $k + \ell$ even. Let (e_α) be an orthonormal basis for $g_{k,\ell}$ (cf. §7.1). A generic element of the basis of $C\ell(k,\ell)$ is

$$c = e_{\alpha_1} \dots e_{\alpha_p} e_{\beta_1} \dots e_{\beta_q} \tag{7.55}$$

$$\text{where } 1 \le \alpha_1 < \dots < \alpha_p \le k < \beta_1 < \dots < \beta_q \le k + \ell \tag{7.56}$$

and it is understood that $c = 1$ for an empty sequence (7.56), i.e. for $p = q = 0$. Therefore, if $p+q > 0$, then

$$\operatorname{Tr} \gamma(c) = 0 \tag{7.57}$$

because, for p+q even, $\gamma(c) = \gamma_{\alpha_1} \gamma_{\alpha_2} \cdots \gamma_{\beta_q} = -\gamma_{\alpha_2} \cdots \gamma_{\beta_q} \gamma_{\alpha_1}$, and Tr MN = Tr NM for every M, N $\in$ End S; for p+q odd $\gamma(c) = -\Gamma\, \gamma(c)\, \Gamma^{-1}$ also leads to (7.57). The basis consisting of $2^{k+\ell}$ elements (7.55) is orthogonal with respect to both h and h'. To compute the signature of h and h' we note that

$$h(c,c) = (-1)^q \quad \text{and} \quad h'(c,c) = (-1)^p$$

so that h is positive definite for $\ell = 0$ and h' is positive-definite for $k = 0$. Given p and q, there are $\binom{k}{p}\binom{\ell}{q}$ different elements of the form (7.55). Therefore, putting

$$w(x, y) = (1+x)^k (1+y)^\ell = \sum_{p=0}^{k} \sum_{q=0}^{\ell} \binom{k}{p}\binom{\ell}{q} x^p y^q$$

we see that

$$\begin{aligned} \text{index } h &= w(1,-1) = 0 \quad \text{for } \ell > 0, \\ \text{index } h' &= w(-1,1) = 0 \quad \text{for } k > 0. \end{aligned} \tag{7.58}$$

This completes the proof.

By reference to Proposition 4.2 and (7.38) one could now prove

Theorem 7.4 The Hermitean forms A and D defined on the carrier space S of an irreducible representation γ of $C\ell(k,\ell)$ are such that

$$\begin{aligned} &\text{A is definite for } k > 0 \text{ and } \ell = 0, \\ &\text{D is definite for } k = 0 \text{ and } \ell > 0, \end{aligned} \tag{7.59}$$

and they are neutral otherwise whenever defined.

The importance of this theorem justifies giving it another, direct *proof*. This can be done by making appeal to concrete representations of the algebras. We choose the Cartan representation described in §6.3 and adapt it to real algebras as follows.

Let $k + \ell = 2m$ and γ_α ($\alpha = 1, \ldots, 2m$) be the matrices (6.26). If $k \geq \ell$ then we put

$$\begin{aligned} \gamma'_{2\alpha-1} &= \gamma_{2\alpha-1} && \text{for } \alpha = 1, \ldots, m \\ \gamma'_{2\alpha} &= i\,\gamma_{2\alpha} && \text{for } \alpha = 1, \ldots, (k-\ell)/2 \\ \gamma'_{2\alpha} &= \gamma_{2\alpha} && \text{for } \alpha = 1 + (k-\ell)/2, \ldots, m \end{aligned} \tag{7.60}$$

so that the matrices γ'_α ($\alpha = 1, \ldots, 2m$) generate a representation of $C\ell(k,\ell)$. The intertwining matrices A and D,

$$\gamma'^{\dagger}_\alpha = A\,\gamma'_\alpha A^{-1} = -D\,\gamma'_\alpha D^{-1}$$

are given, up to phase factors, by

$$\gamma_{k-\ell+2}\ \gamma_{k-\ell+4} \cdots \gamma_{2m} \sim A \text{ (for } \ell \text{ even) and } D \text{ (for } \ell \text{ odd)},$$

$$\gamma_{k-\ell+2}\ \gamma_{k-\ell+4} \cdots \gamma_{2m}\Gamma \sim A \text{ (for } \ell \text{ odd) and } D \text{ (for } \ell \text{ even)},$$

with the understanding that the product of an empty sequence of matrices is the unit matrix. Only in the latter case is the corresponding Hermitean form definite. For $k \geq \ell$ this can happen only for $k = 2m$ and $\ell = 0$: then $A = I$ and $D \sim \Gamma$. The case $k < \ell$ is considered in a similar manner,only D can then be definite and this happens for $k = 0$ and $\ell = 2m$.

The matrices γ'_α given by (7.60) are all Hermitian for ℓ=0, the matrix

$$C = \begin{cases} \gamma_2\gamma_4\cdots\gamma_{2m} & \text{for m even,} \\ \gamma_1\gamma_3\cdots\gamma_{2m-1} & \text{for m odd,} \end{cases}$$

is real and

$$C^{-1} = {}^tC = (-1)^{m(m-1)/2}C. \tag{7.61}$$

If

$$\gamma'^{\dagger}_\alpha = \gamma'_\alpha,\ \bar{\gamma}'_\alpha = C\,\gamma'_\alpha\, C^{-1},\ {}^tC = C = \bar{C} = C^{-1}$$

then the matrices

$$'\gamma_\alpha = U^{-1}\,\gamma'_\alpha\, U \tag{7.62}$$

where U is the unitary matrix

$$U = \tfrac{1}{2}(1+i)I + \tfrac{1}{2}(1-i)C$$

are real, $'\bar{\gamma}_\alpha = {}'\gamma_\alpha$, and also Hermitean, therefore symmetric. By virtue of (7.61), these conditions are satisfied in signature (2m,0) where m=0 or 1 mod 4. If m=0 mod 4, then the product $'\gamma_1 .. '\gamma_{2m}$ is also real symmetric and its square is I.

In signature (0,2m), the matrices γ'_α given by (7.60) are all anti-Hermitean. A similar reasoning, with C replaced by CΓ, shows that for m=0 or 3 mod 4 one can transform γ'_α into matrices $'\gamma_\alpha$ which are real and skew. For m=3 mod 4 the product $'\gamma_1 .. '\gamma_{2m}$ is also real skew and its square is -I. If M is an anti-Hermitean N by N matrix, $M \in \mathbb{C}(N)$, $M^\dagger = -M$, then its real form is skew; explicitly, if M = P+iQ, where $P, Q \in \mathbb{R}(N)$, then the real form of M is the skew, 2N by 2N matrix

$$\begin{pmatrix} P & -Q \\ Q & P \end{pmatrix}$$

Therefore, the algebras $C\ell(0,2m)$ for m ≡ 1 and 2 mod 4 and the algebras $C\ell(0,2m+1)$ for m ≡ 0, 1 and 2 mod 4 admit real representations of dimension 2^{m+1} such that the corresponding Dirac matrices are skew. This can be summarized in

Theorem 7.5 For every integer n>0, the Clifford algebra $C\ell(0,n)$ admits an irreducible representation in a real spinor space of dimension $2^{\chi(n)}$, where χ is the *Radon-Hurwitz sequence* defined by

n =	1	2	3	4	5	6	7	8
χ(n) =	1	2	2	3	3	3	3	4

and

$$\chi(n+8) = \chi(n) + 4.$$

The representation can be chosen is such a way that the real Dirac matrices are *skew*. For n even and n≡1 mod 4, the rerpesentation is faithful; for n≡3 mod 4, the restriction of the representation to the even subalgebra is faithful and irreducible.

Let

$$\gamma : C\ell(0,n) \to \mathbb{R}(2^{\chi(n)}) \tag{7.63}$$

be a representation described in Theorem 7.5. Let $t \in \mathbb{R}$ and $u = (u^\alpha) \in \mathbb{R}^n$ be a vector, then the matrix

$$\gamma(t+u) = tI + u^{\alpha}\gamma_{\alpha}$$

is invertible unless t=0 and u=0: indeed, we have

$$\gamma(t+u)\gamma(t-u) = (t^2 + \sum_{\alpha=1}^{n} (u^{\alpha})^2)I.$$

Let φ be a real, $2^{\chi(n)}$-component spinor; the representation γ defines an orthogonal multiplication

$$\mathbb{R}^{n+1} \times \mathbb{R}^{2^{\chi(n)}} \rightarrow \mathbb{R}^{2^{\chi(n)}}, \tag{7.64a}$$

$$(t+u) \cdot \varphi = \gamma(t+u)\varphi. \tag{7.64b}$$

For n=1, 3 and 7— and only for these positive integeres — we have

$$n+1 = 2^{\chi(n)}$$

so that (7.64) is a multiplication in the vector space $\mathbb{R}^{n+1}$ which makes it into a division algebra; the cases n = 1,3 and 7 correspond to complex numbers, quaternions and octonions (Cayley numbers), respectively. In the last case the multiplication is non-associative.

The representation (7.63) can be also used to construct sets of vector fields on spheres which are, at each point of the sphere, linearly independent and orthogonal. It is well-known that even-dimensional spheres do not admit any nowhere vanishing vector fields. Let N be an even integer and n *the largest integer such that*

$$N = 2^{\chi(n)}\, p,$$

where p is odd. Consider the unit, (N-1)-dimensional sphere

$$S_{N-1} = \{\varphi \in \mathbb{R}^N \mid {}^t\varphi\varphi = 1\}.$$

Let γ' be the (decomposable) representation

$$\gamma' : C\ell(0,n) \rightarrow \mathbb{R}(N) = \mathbb{R}(2^{\chi(n)}) \otimes \mathbb{R}(p)$$

defined by

$$\gamma'(a) = \gamma(a)\otimes I, \text{ where } a\in C\ell(0,n).$$

Since the matrices $\gamma(u)$, $u \in \mathbb{R}^n$, are skew, so are the N by N matrices $\gamma'(u)$. Therefore, for every $u \in \mathbb{R}^n$ and $\varphi \in S_{N-1}$ the vector $\gamma'(u)\varphi$ is orthogonal to φ, i.e. tangent to S_{N-1}. Taking for u the vectors $e_1, \ldots, e_n$ of an orthonormal basis in $\mathbb{R}^n$ we obtain a set $\gamma'(e_\alpha)\varphi$ ($\alpha=1,\ldots,n$) of n orthonormal vector fields tangent to S_{N-1}. Adams (1962) showed that no more everywhere linearly independent vector fields can be constructed on any sphere: the Clifford construction provides the best one can do. Moreover, since $n+1 < 2^{\chi(n)}$ unless n=1, 3 or 7, the spheres S_1, S_3 and S_7 are the only ones admitting "teleparallelism": their tangent bundles are trivial.

7.5 The spinorial chessboard

There are several "periodicity properties" of real Clifford algebras and their representations. The class $[C\ell(k,\ell)]$ of the algebra, cf. § 7.1, depends only on $k-\ell$ mod 8. But the symmetry properties of the invariant bilinear forms depend on $k+\ell$ mod 8. There is a "double periodicity" in the set of all real Clifford algebras: it is convenient to describe it by referring it to a chessboard.

We define the *spinorial chessboard* to be the set of 64 real algebras

$$\{\, C\ell(k,\ell) \mid 0 \le k,\ell \le 7 \}$$

where it is understood that $C\ell_0(0,0) \to C\ell(0,0)$ is the algebra $\mathbb{R} \to \mathbb{R}$, i.e. $C\ell_1(0,0) = \{0\}$.In addition to the chessboard — and representations of its elements — we consider the two eight-dimensional Euclidean algebras $C\ell(8,0)$ and $C\ell(0,8)$. According to the periodicity property (7.11), if $k' = k+8p$ and $\ell' = \ell+8q$, then

$$C\ell(k',\ell') = C\ell(k,\ell) \otimes \mathbb{R}(16^{p+q}). \tag{7.65}$$

Therefore, every Clifford algebra can be represented as in (7.65), with $C\ell(k,\ell)$ on the chessboard. The significance of this remark goes beyond the mere isomorphism of algebras (7.65): the representations of $C\ell(k',\ell')$ and the associated bilinear and Hermitean forms can be easily constructed from those of $C\ell(k,\ell)$. Adding eight dimensions makes larger the Clifford algebra and the associated spinor spaces, but preserves their essential properties such as the symmetry of B, type of C, etc.

To make the last statement more precise, consider a vector space $V=\mathbb{R}^8$ with a positive-definite scalar product. The faithful irreducible representation of its Clifford algebra,

$$C\ell(8,0) \to \mathrm{End}S, \tag{7.66}$$

is real so that S can be taken to be a real, 16-dimensional space (of Majorana spinors). Let $(e_1,...,e_8)$ be an orthonormal basis in V. The set of 2^8 products of the form (5.14) is a basis of the algebra. This basis is orthogonal for the scalar product h on $C\ell(8,0)$ defined by (4.20): if

$$a = e_{\alpha_1} \dots e_{\alpha_p} \text{ and } b = e_{\beta_1} \dots e_{\beta_q},$$

where

$$1 \leq \alpha_1 < \dots < \alpha_p \leq 8 \text{ and } 1 \leq \beta_1 < \dots < \beta_q \leq 8,$$

then

$$\beta(a)b = 1 \text{ whenever } p = q \text{ and } \alpha_1 = \beta_1, \dots, \alpha_p = \beta_p,$$

and

$$\operatorname{Tr} \gamma(\beta(a)b) = 0 \text{ otherwise.}$$

Therefore, the scalar product h is positive-definite and, by Proposition 4.2, the symmetric bilinear form B is also positive-definite. We choose a basis in S such that B is represented by a unit matrix with respect to this basis, and we use the basis to identify S with $\mathbb{R}^{16}$ so that the representation (7.66) can be described as

$$\theta : C\ell(8,0) \to \mathbb{R}(16) \tag{7.67}$$

and $\check{\theta} = \theta$, i.e. the Dirac matrices

$$\theta_\alpha = \theta(e_\alpha), \quad \alpha = 1,\dots,8,$$

are symmetric,

$$^t\theta_\alpha = \theta_\alpha.$$

The image of the volume element by θ,

$$\Theta = \theta_1 \dots \theta_8,$$

is also symmetric and $\Theta^2 = I$. There is the decomposition

$$\theta_0 = \theta_+ \oplus \theta_-,$$

where

$$\theta_\pm : C\ell_0(8,0) \to \mathbb{R}(8) \tag{7.68}$$

are the inequivalent Weyl representations of the even algebra. Since Θ anticommutes with the Dirac matrices, one can construct a faithful irreducible representation of the opposite algebra

$$*\theta : C\ell(0,8) \to \mathbb{R}(16) \tag{7.67*}$$

by putting

$$*\theta_\alpha = \Theta\,\theta_\alpha, \quad \alpha = 1, \ldots, 8, \tag{7.69}$$

so that the Dirac matrices (7.69) are skew and

$$^t_*\theta_\alpha = \Theta\ {}_*\theta_\alpha\,\Theta^{-1}. \tag{7.70}$$

Let

$$\gamma : C\ell(k,\ell) \to \mathrm{End}S \tag{7.71}$$

be a representation of the Clifford algebra $C\ell(k,\ell)$. One can extend it to representations

$$\gamma' : C\ell(k+8,\ell) \to \mathbb{R}(16) \otimes \mathrm{End}\ S$$

and

$$\gamma'' : C\ell(k,\ell+8) \to \mathbb{R}(16) \otimes \mathrm{End}\ S$$

by putting

$$\gamma'_\alpha = \Theta \otimes \gamma_\alpha = \gamma''_\alpha \ (\alpha = 1, \ldots, k+\ell), \tag{7.72a}$$

$$\gamma'_{\alpha+k+\ell} = \theta_\alpha \otimes \mathrm{I} \quad (\alpha = 1, \ldots, 8), \tag{7.72b}$$

and

$$\gamma''_{\alpha+k+\ell} = \Theta\,\theta_\alpha \otimes \mathrm{I} \quad (\alpha = 1, \ldots, 8). \tag{7.72c}$$

Marking with primes or double primes the quantities corresponding to the extensions γ' or γ'', respectively, we obtain for $k+\ell$ even

$$\Gamma' = \Theta \otimes \Gamma = \Gamma''$$

$$\begin{aligned} &A' = I \otimes A, \quad A'' = \Theta \otimes A \\ &B' = I \otimes B, \quad B'' = \Theta \otimes B \\ &C' = I \otimes C = C'' \\ &D' = \Theta \otimes D, \quad D'' = I \otimes D \\ &E' = \Theta \otimes E, \quad E'' = I \otimes E. \end{aligned} \tag{7.73}$$

Adding 8 "positive" or "negative" dimensions preserves the character of A,B,C or D,E,C, respectively. If A or D is definite, then so is A' or D'', respectively. There are similar results for $k+\ell$ odd, see § 7.6.5.

Table VI

The Spinorial Chessboard

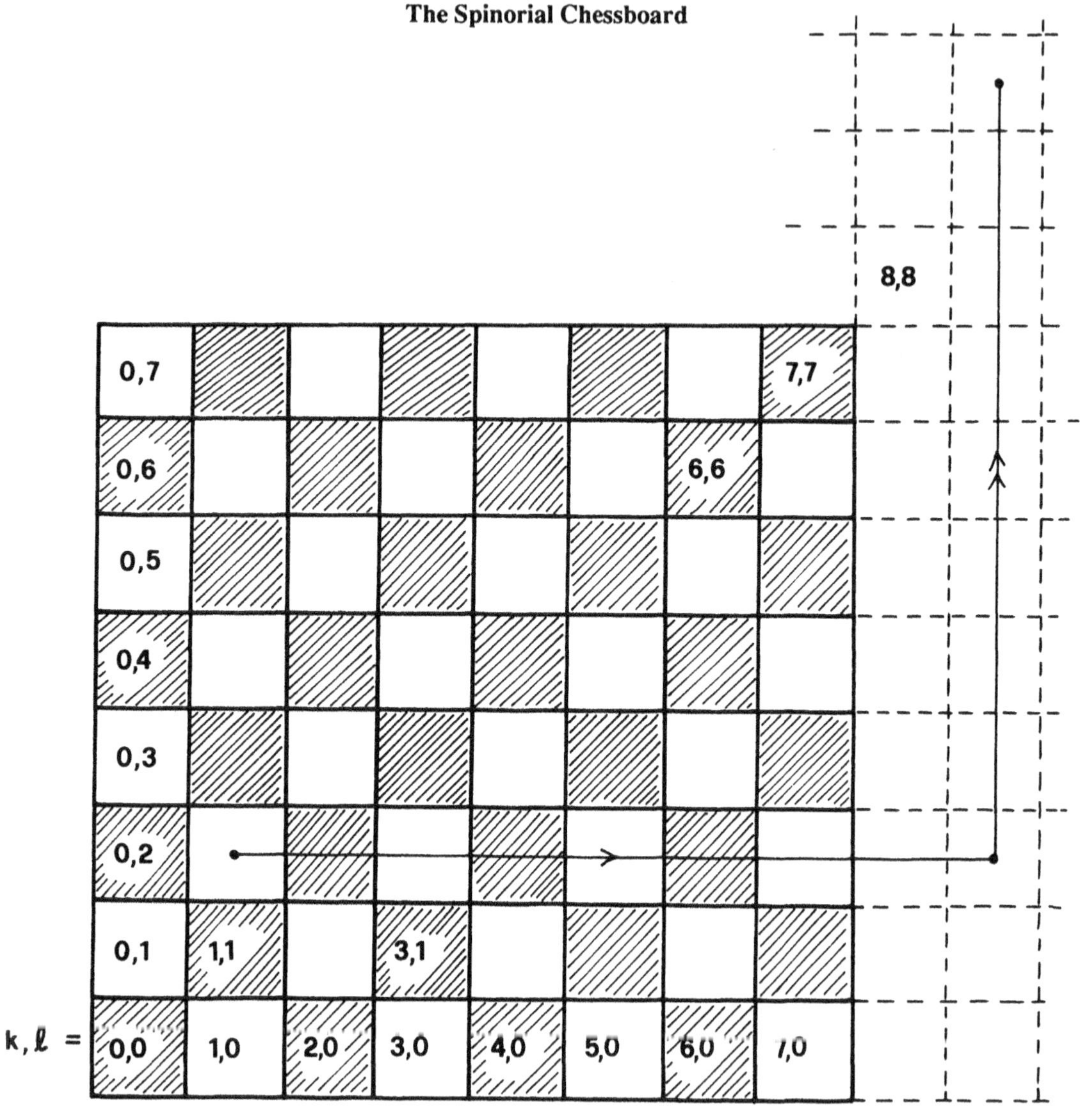

Even- and odd-dimensional Clifford algebras $C\ell(k, \ell)$, $0 \leq k, \ell \leq 7$, occupy, respectively, black and white squares of the board. For example, the algebra $C\ell(3,1)$ of Minkowski space is at the square of the white queen's pawn. Every real Clifford algebra can be reached from one on the board with rook's moves to the right and upwards, each move being by a multiple of eight squares, as described by (7.72) and (7.73).

Table VII

The structure of the algebras occurring on the chessboard may be determined from the following data:

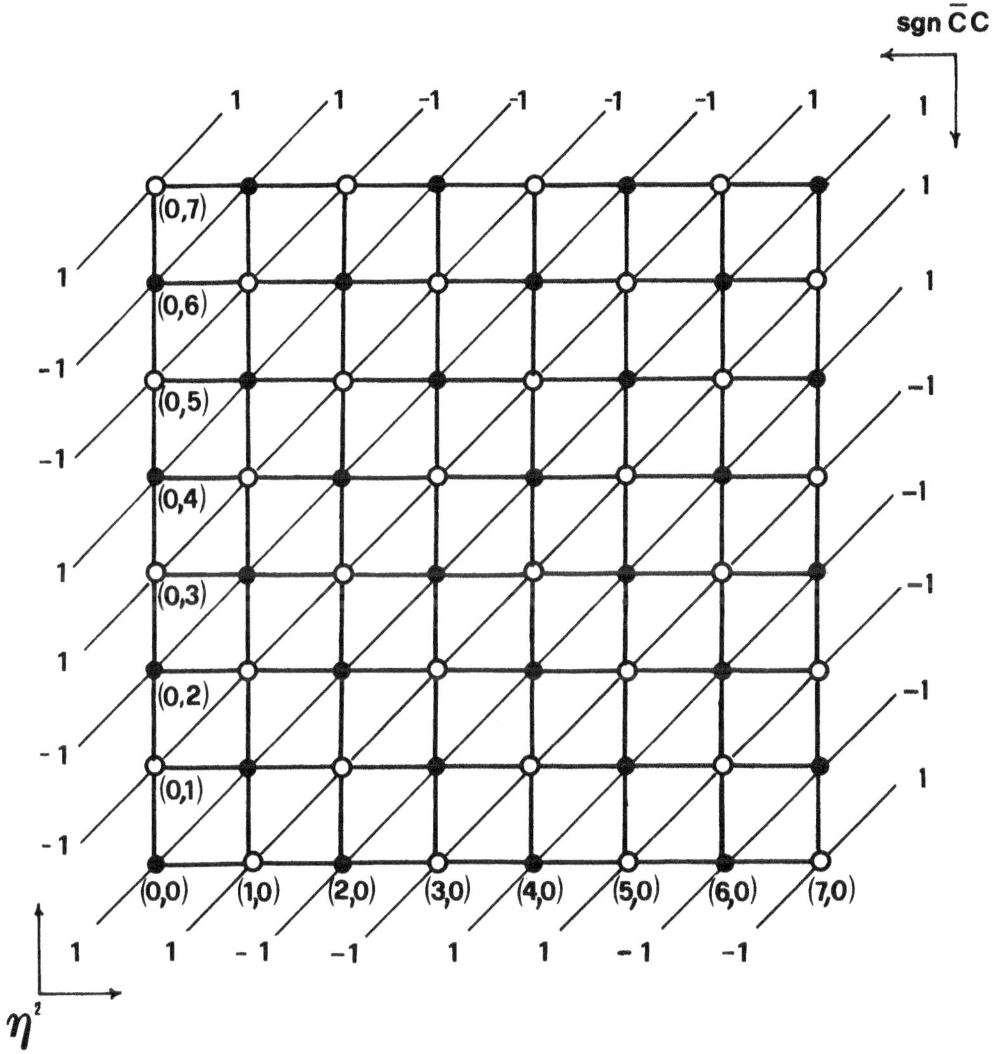

White and black dots replace here the squares of the chessboard. The figures on the left and lower sides are values of the volume element squared. Those on the right and upper sides determine the type (real if 1, quaternionic if -1) of the full (for k + ℓ even) or even (for k + ℓ odd) Clifford algebra.

Table VIII

The bilinear forms and their symmetries

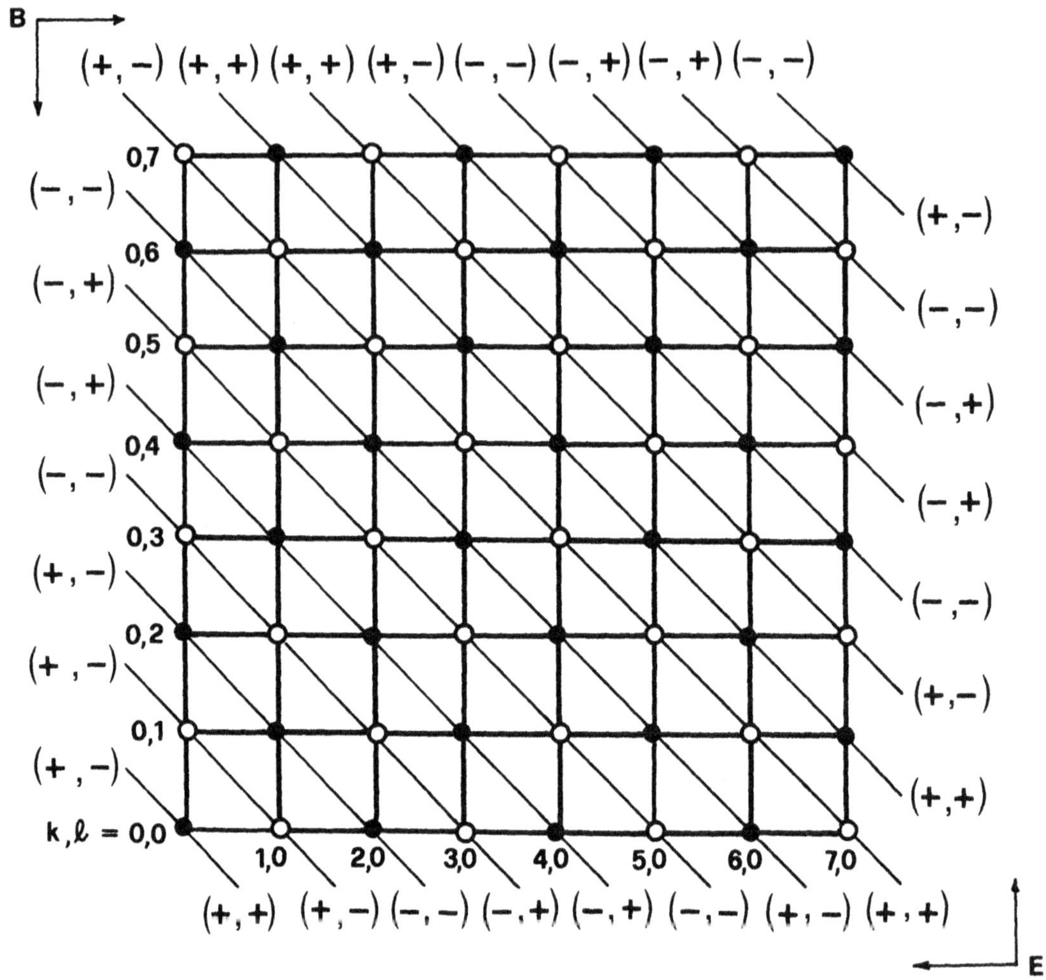

The isomorphisms B and E defined by ${}^t\gamma_\alpha = B\,\gamma_\alpha\,B^{-1}$ and ${}^t\gamma_\alpha = -E\gamma_\alpha\,E^{-1}$ are either symmetric or skew and they either commute or anticommute with the helicity operator Γ. These properties are indicated above by pairs $(\varepsilon_1, \varepsilon_2)$ where ε_1 and ε_2 = + or -. They are defined by ${}^tB = \varepsilon_1\,B$ and $B\Gamma = \varepsilon_2\,{}^t\Gamma B$; and similarly for E.

Table IX

The Dirac (Hermitean) forms

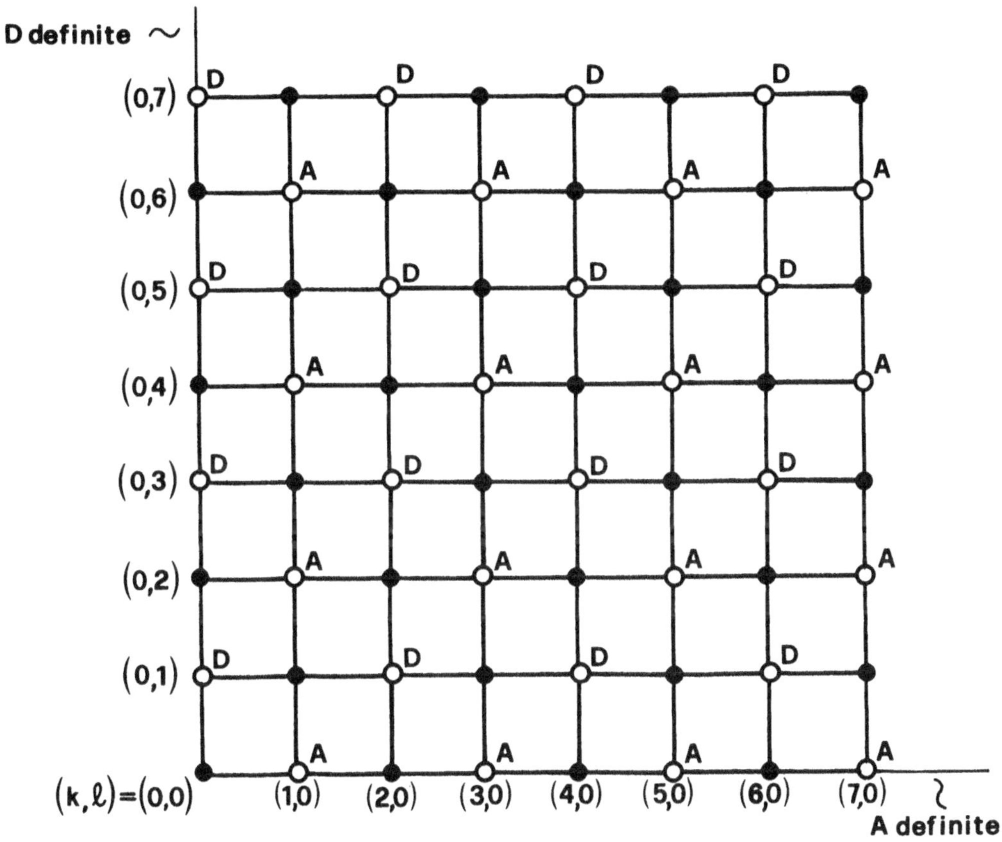

The isomorphisms A and D are defined by $\gamma^\dagger_\alpha = A\,\gamma_\alpha\,A^{-1}$ and $\gamma^\dagger_\alpha = -D\,\gamma_\alpha\,D^{-1}$. They both exist for even dimensional spaces. In an odd number of dimensions, exactly one of the two exists, depending on the parity of k; this is indicated by the letter A or D next to the corresponding white dot. The Hermitean forms A(ϕ, ϕ) are (positive) definite for the algebras $C\ell(k,0)$; similarly, the Hermitean forms D(ϕ, ϕ) are (positive) definite for $C\ell(0,\ell)$. Otherwise they are neutral.

7.6 Summary

In this section we give a short summary of the properties of representations of Clifford algebras of real vector spaces in a language familiar to physicists. The 2^m-dimensional spinor space S is identified with $\mathbb{C}^{2^m}$, the endomorphsims γ_α are 2^m by 2^m matrices and the symbols ${}^tA$, $A^\dagger$ and $\bar{A}$ denote the usual transpose, Hermitean conjugate and complex conjugate of the matrix A, respectively. Therefore $A^\dagger = {}^t\bar{A}$.

If (k,ℓ) is the signature, $k+\ell = 2m$ or $2m+1$, then there are $k+\ell$ Dirac matrices $\gamma_\alpha \in \mathbb{C}(2^m)$ such that

$$\gamma_\alpha \gamma_\beta + \gamma_\beta \gamma_\alpha = 0 \quad \text{for} \quad \alpha \neq \beta, \ \alpha \ \text{and} \ \beta = 1, \ldots, k+\ell, \tag{7.74a}$$

$$\gamma_\alpha^2 = I \text{ for k values of } \alpha \text{ and } \gamma_\alpha^2 = -I \text{ for } \ell \text{ values of } \alpha . \tag{7.74b}$$

We do not insist here that the first k values of the label should correspond to Dirac matrices with positive squares; only the total numbers of positive and negative squares matter.

7.6.1. The case of even-dimensional spaces, $k+\ell = 2m$.

Let $k-\ell = 8p+\nu$, where p is an integer and $0 \leq \nu \leq 7$. The matrix

$$\Gamma = i^{\nu(\nu-1)/2}\, \gamma_1 \ldots \gamma_{2m} \text{ anticommutes with } \gamma_\alpha, \tag{7.75}$$

and

$$\Gamma^2 = I. \tag{7.76}$$

There exist invertible matrices A, B, C, D, E $\in \mathbb{C}(2^m)$ such that for every α

$$\gamma^\dagger_\alpha = A\gamma_\alpha A^{-1}, \tag{7.77A}$$

$${}^t\gamma_\alpha = B\gamma_\alpha B^{-1}, \tag{7.77B}$$

$$\bar{\gamma}_\alpha = C\gamma_\alpha C^{-1}, \tag{7.77C}$$

$$\gamma_\alpha^\dagger = -D\gamma_\alpha D^{-1}, \tag{7.77D}$$

$${}^t\gamma_\alpha = -E\gamma_\alpha E^{-1}. \tag{7.77E}$$

They satisfy

$$^{t}B = (-1)^{m(m-1)/2}B \tag{7.78B}$$

$$^{t}E = (-1)^{m(m+1)/2}E \tag{7.78E}$$

$$^{t}\Gamma = (-1)^{m}B\Gamma B^{-1} \tag{7.78Γ}$$

The defining properties (7.77) determine the matrices A, ..., E up to complex factors. These factors can be chosen so that

$$\bar{C}C = (-1)^{\nu(\nu-2)/8}\, I \tag{7.79}$$

$$A = \bar{B}C = A^{\dagger} \tag{7.80A}$$

$$D = \bar{E}C = D^{\dagger} \tag{7.80D}$$

$$E = i^{\ell}B\Gamma \tag{7.81}$$

The remaining freedom is $A \to \lambda A$, $B \to \lambda\mu B$, $C \to \mu C$, $D \to \lambda D$, $E \to \lambda\mu E$, where λ is real $\neq 0$ and μ is complex of unit modulus.

If U is an invertible matrix, $U \in \mathbb{C}(2^m)$, then the matrices

$$'\gamma_\alpha = U^{-1}\gamma_\alpha U \tag{7.82}$$

have the properties (7.74). Marking with primes on the left the matrices associated by (7.77A-E) with the matrices $'\gamma_\alpha$, we have

$$'A = U^{\dagger}AU, \tag{7.83A}$$

$$'B = {}^{t}UBU, \tag{7.83B}$$

$$'C = \bar{U}^{-1}CU, \tag{7.83C}$$

and similar relations for $'\Gamma$, $'D$ and $'E$.

The Hermitean forms $\varphi^{\dagger}A\varphi$ and $\varphi^{\dagger}D\varphi$, where $\varphi \in \mathbb{C}^{2^m}$, are *neutral* except in the following cases:

$$\varphi^\dagger A \varphi \text{ is } \textit{definite} \text{ for } \ell = 0, k > 0, \tag{7.84A}$$

$$\varphi^\dagger D \varphi \text{ is } \textit{definite} \text{ for } k = 0, \ell > 0. \tag{7.84D}$$

These forms restrict to non-degenerate Hermitean forms on the spaces of Weyl spinors if, and only if, k is even. For odd k, the matrices A and D change the helicity of Weyl spinors.

7.6.2 The case of odd-dimensional spaces, $k+\ell = 2m+1$

Let $k-\ell = 8p+\nu$, where p is an integer and $1 \leq \nu \leq 7$. One can choose the matrices $\gamma_1,\ldots, \gamma_{2m+1}$ so that

$$\gamma_1 \cdots \gamma_{2m+1} = i^{\nu(\nu-1)/2}\, I. \tag{7.85}$$

There exist matrices A_o, B_o and C_o such that, for every α

$$\gamma^\dagger{}_\alpha = (-1)^\ell A_o \gamma_\alpha A^{-1}{}_o \tag{7.86A}$$

$$^t\gamma_\alpha = (-1)^m B_o \gamma_\alpha B^{-1}{}_o \tag{7.86B}$$

$$\bar{\gamma}_\alpha = (-1)^{\nu(\nu-1)/2} C_o \gamma_\alpha C^{-1}{}_o \tag{7.86C}$$

and

$$A_o = \bar{B}_o C_o = A_o{}^\dagger \tag{7.87}$$

$$^t B_o = (-1)^{m(m+1)/2} B_o \tag{7.88}$$

$$\bar{C}_o C_o = (-1)^{(\nu^2-1)/8}\, I \tag{7.89}$$

The Hermitean form $\varphi^\dagger A_o \varphi$ is *neutral* except in the case when either $k = 0$ or $\ell = 0$: it is then *definite*.

7.6.3 Adding one dimension to an even-dimensional space

Let $k+\ell = 2m$ and $k-\ell = 8p+\nu$, as before. The 2m+1 matrices

$$\gamma_1,\ldots,\gamma_{2m} \text{ and } \gamma_{2m+1} = \Gamma \tag{7.90}$$

are Dirac matrices for a space with signature $(k+1,\ell)$ and

$$A_o = \begin{cases} A & \text{for } \ell \text{ even,} \\ D & \text{for } \ell \text{ odd,} \end{cases} \tag{7.91A}$$

$$B_o = \begin{cases} B & \text{for } m \text{ even,} \\ E & \text{for } m \text{ odd,} \end{cases} \tag{7.91B}$$

$$C_o = \begin{cases} C & \text{for } \nu = 0 \text{ or } 4, \\ C\Gamma & \text{for } \nu = 2 \text{ or } 6, \end{cases} \tag{7.91C+}$$

where the matrices Γ, A, ..., E are as in § 7.6.1.

Similarly, the 2m+1 matrices

$$\gamma_1, \ldots, \gamma_{2m} \text{ and } \gamma_{2m+1} = i\Gamma$$

are Dirac matrices for a space with signature $(k,\ell+1)$. The intertwining matrices A_o and B_o are as in (7.91A) and (7.91B), but

$$C_o = \begin{cases} C & \text{for } \nu = 2 \text{ or } 6, \\ C\Gamma & \text{for } \nu = 0 \text{ or } 4. \end{cases} \tag{7.91C-}$$

7.6.4 Adding a 2-dimensional neutral space

As an example, we give explicitly all relevant quantities for an extension from signature (k,ℓ) to $(k+1, \ell+1)$. We choose an extension of type 0 because it is the only one that allows a simultaneous treatment of even- and odd-dimensional spaces. One can take

$$\gamma'_\alpha = \sigma \otimes \gamma_\alpha \ (\alpha = 1, \ldots, k+\ell), \quad \gamma'_{k+\ell+1} = \tau \otimes I \quad \text{and} \ \gamma'_{k+\ell+2} = \varepsilon \otimes I \tag{7.92}$$

(i) For $k+\ell = 2m$ we have

$$\begin{aligned} \Gamma' &= \sigma \otimes \Gamma, \quad C' = I \otimes C \\ A' &= \tau \otimes D, \quad D' = (-1)^\ell \, i\, \varepsilon \otimes A \\ B' &= \tau \otimes E, \quad E' = (-1)^{\ell+1} \, i\, \varepsilon \otimes B \end{aligned} \tag{7.93}$$

(ii) For $k+\ell = 2m+1$ we have

$$A'_o = \begin{cases} i\,\varepsilon \otimes A_o & \text{for } \ell \text{ even,} \\ \tau \otimes A_o & \text{for } \ell \text{ odd,} \end{cases} \tag{7.94A}$$

$$B'_o = \begin{cases} -i\,\varepsilon \otimes B_o & \text{for } m \text{ even,} \\ \tau \otimes B_o & \text{for } m \text{ odd,} \end{cases} \tag{7.94B}$$

$$C'_o = \begin{cases} I \otimes C_o & \text{for } \nu = 1 \text{ or } 5, \\ i\sigma \otimes C_o & \text{for } \nu = 3 \text{ or } 7, \end{cases} \tag{7.94C}$$

where $k-\ell = 8p+\nu$ and the matrices A'_o, B'_o and C'_o are in the same relation to γ'_α as the matrices A_o, B_o and C_o are to γ_α, cf. § 7.6.2.

7.6.5 Adding an 8-dimensional Euclidean space

The eight 16 by 16 real matrices

$$\begin{aligned} \theta_1 &= \sigma \otimes I \otimes I \otimes I, & \theta_2 &= \varepsilon \otimes \varepsilon \otimes I \otimes I, \\ \theta_3 &= \varepsilon \otimes \sigma \otimes \varepsilon \otimes I, & \theta_4 &= \varepsilon \otimes \sigma \otimes \sigma \otimes \varepsilon, \\ \theta_5 &= \varepsilon \otimes \sigma \otimes \tau \otimes \varepsilon, & \theta_6 &= \varepsilon \otimes \tau \otimes I \otimes \varepsilon, \\ \theta_7 &= \varepsilon \otimes \tau \otimes \varepsilon \otimes \sigma, & \theta_8 &= \varepsilon \otimes \tau \otimes \varepsilon \otimes \tau, \end{aligned} \tag{7.95}$$

are symmetric, satisfy

$$\theta_\alpha \theta_\beta + \theta_\beta \theta_\alpha = 2\delta_{\alpha\beta} \quad (\alpha,\beta = 1,\ldots,8);$$

their product

$$\Theta = \tau \otimes I \otimes I \otimes I$$

is also symmetric and $\Theta^2 = I$.

From a representation of $C\ell(k,\ell)$ generated by the Dirac matrices γ_α (α=1, ..., k+ℓ) one can construct the representation γ' and γ'' of $C\ell(k+8,\ell)$ and $C\ell(k,\ell+8)$, respectively. They are given by (7.72). If $k+\ell = 2m$ or $2m+1$, then the matrices γ' and γ'' are 2^{m+4} by 2^{m+4}. Marking with primes or double primes the quantities corresponding to the extensions γ' or γ'', respectively, for $k+\ell$ even we have formulae (7.73) for the intertwining isomorphisms, whereas for $k+\ell = 2m+1$,

$$A'_0 \text{ (for } \ell \text{ even) and } A''_0 \text{ (for } \ell \text{ odd)} = I \otimes A_0, \tag{7.96A}$$

$$B'_0 \text{ (for } m \text{ even) and } B''_0 \text{ (for m odd)} = I \otimes B_0, \tag{7.96B}$$

$$C'_0 \text{ and } C''_0 \text{ (for } \nu = 1 \text{ or } 5) = I \otimes C_0, \tag{7.96C}$$

$$A'_0 \text{ (for } \ell \text{ odd) and } A''_0 \text{ (for } \ell \text{ even)} = \Theta \otimes A_0, \tag{7.97A}$$

$$B'_0 \text{ (for } m \text{ odd) and } B''_0 \text{ (for m even)} = \Theta \otimes B_0, \tag{7.97B}$$

$$C'_0 \text{ and } C''_0 \text{ (for } \nu = 3 \text{ or } 7) = \Theta \otimes C_0. \tag{7.97C}$$

Table X

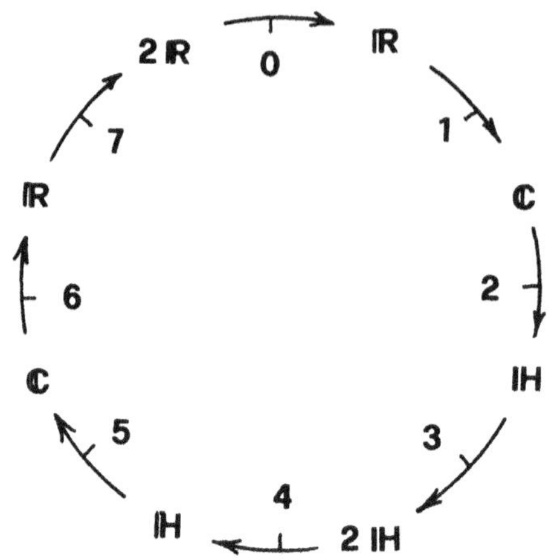

The real clock*

may be used to find the Clifford algebra $C\ell(k, \ell)$ and its even subalgebra $C\ell_0(k, \ell)$: compute first the *hour* μ such that $\ell - k = 8p + \mu$, where p is an integer and $0 \leq \mu \leq 7$. The letters adjacent to the hour determine the type of the algebras. The dimension of the full algebra is $2^{k+\ell}$. For example, $C\ell_0(3,5) \to C\ell(3,5)$ is $\mathbb{C}(8) \to H(8)$ because, in this case, $\mu = 2$ and $\dim H(8) = 2^8$.

---

* The complex clock is much simpler: it has a two-hour dial.

REFERENCES

Artin E 1957 *Geometric algebra* (New York: Interscience)

Atiyah M F, Bott R and Shapiro A 1964 *Clifford modules* Topology **3** Suppl. **1** 3-38

Atiyah M F and Singer I M 1968 *The index of elliptic operators. III* Ann. of Math. **87** 546-604

Benn I M and Tucker R M 1984 *Pure spinors and real Clifford algebras.* preprint (1984)

— *An introduction to spinors and geometry with applications in physics.* To appear A. Hilger (1988)

Borel A and Hirzebruch F 1958, 1959 and 1960 *Characteristic classes and homogeneous spaces* Amer. J. Math **80** 458-538, **81** 315-382 and **82** 491-504

Bourbaki N 1959 *Algèbre* Ch. 9 § 4 and 9 (Paris : Hermann)

— 1982 *Groupes et algèbres de Lie* Ch. 9 (Paris : Masson)

Brauer R and Weyl H 1935 *Spinors in n dimensions* Amer. J. Math. **57** 425-449

Budinich P 1979 *On conformally covariant field equations* Czech. J. Phys. **B 29** 6-21

— 1984 *On pure, conformal and exotic spinors* Proc. Conf. on Group Theoretical Methods in Physics, Trieste 1983, Lecture Notes in Physics (Berlin : Springer)

— 1986a *Null vectors, spinors and strings* Comm. Math. Phys. **107** 455-465

— 1986b *Pure spinors and quadric Grassmanians* Phys. Rep. **137** 35-47

— and Furlan P 1982 and 1983 *On Dirac-like equations in 2n-dimensional spaces* Nuovo Cimento **A 70** 243-272 and **A 76** 569-595

— and Rigoli M 1987 *Cartan spinors, minimal surfaces, and strings* SISSA (Trieste) preprint 95/87/F.M.

— and Trautman A 1986 *Remarks on pure spinors* Lett. Math. Phys. **11** 315-324

Cartan E 1908 *Nombres complexes* Encycl. Sc. Math. **I** 5 éd. fran. 329-468; Oeuvres Complètes Partie II, vol. 1, 107-246 (Paris 1953 : Gauthier-Villars)

— 1913 *Sur les groupes projectifs qui ne laissent invariante aucune multiplicité plane* Bull. Soc. Math. France **41** 53-96

— 1922 *Sur les espaces conformes généralisés et l'Univers optique* C.R. Acad. Sci. Paris **174** 857-859

— 1925 *Le principe de dualité et de la théorie des groupes simples et semi-simples* Bull. Soc. Math. France **49** 361-374

— 1938 *La théorie des spineurs I & II* (Paris : Hermann); English transl. by R Streater *The theory of spinors* (Paris 1966 : Hermann)

Case K M 1955 *Biquadratic spinor identities* Phys. Rev. **97** 810-823

Cayley A 1845 *On certain results relating to quaternions* Phil. Mag. **26** 141-145

— 1855 *Recherches ultérieures sur les déterminants gauches* J. reine und angew. Math. (Crelle) **50** 299-313

Chevalley C 1954 *The algebraic theory of spinors* (New York : Columbia University Press)

Clifford W K 1878 *Applications of Grassmann's extensive algebra* Amer. J. Math. **1** 350-358

Coqueraux R 1985 *Spinors, reflections and Clifford algebras: a review* University of Texas (Austin) preprint UTTG 21/85

Dabrowski L and Trautman A 1986 *Spinor structures on spheres and projective spaces* J. Math. Phys. **27** 2022-2028

Darwin C G 1928 *The wave equation of the electron* Proc. R. Soc. (London) **A 118** 654-680

Dirac P A M 1928 *The quantum theory of the electron I & II* **A 117** 610-624 & **A 118** 351-361

— 1958 *The principles of quantum mechanics* 4th rev. ed. (Oxford : Clarendon Press)

Euler L 1770 Nuovi Comm. Acad. Petrop. **15** 101

Good R H Jr 1955 *Properties of the Dirac matrices* Rev. Mod. Phys. **27** 187-211

Graf W 1978 *Differential forms as spinors* Ann. Inst. H. Poincaré **A 29** 85-109

Gürsey F 1958 *Relation of charge independence and baryon conservation to Pauli's transformation* Nuovo Cimento **7** 411-415

Gutt S 1986 *Killing spinors on spheres and projective spaces* Université Libre (Bruxelles) preprint

Haefliger A 1956 *Sur l'extension du groupe structural d'un espace fibré* C.R. Acad. Sci. Paris **243** 558-560

Helfer A D 1983 *Remarks on pure spinors* Twistor Newsletter **16** 40

Igusa J 1970 *A classification of spinors up to dimension twelve* Amer. J. Math. **92** 997-1028

Ivanenko D and Landau L 1928 *Zur Theorie des magnetischen Elektrons. I* Z. Physik **48** 340-348

Kähler E 1960 *Innerer und äusserrer Differentialkalkül* Abh. Deutsch. Akad. Wiss. Berlin (Math.- Phys.) **4**

Karrer G 1973 *Darstellung von Cliffordbündeln* Ann. Acad. Sci. Fennicae Ser. A I Math. No. 521

Kline M 1972 *Mathematical thought from ancient to modern times* p. 772 (New York : Oxford University Press)

Lipschitz R O 1886 *Untersuchungen ueber die Summen von Quadraten* (Bonn : Max Cohen und Sohn); cf also Bull. Sci. Math. 2 Sér. **10** 163-183

Lipschitz R (signed) 1959 *Correspondence* Ann. of Math. **69** 247-251

Lounesto P 1981 *Scalar products of spinors and an extension of Brauer-Wall groups* Found. of Physics **11** 721-740

Majorana E 1937 *Teoria simmetrica dell'elettrone e del positrone* Nuovo Cimento **14** 171-185

Ne'eman Y 1978 *Spinor-type fields with linear, affine and general coordinate transformations* Ann. Inst. H. Poincaré **A 28** 369-378

Newman E T and Penrose R 1962 *An approach to gravitational radiation by a method of spin coefficients* J. Math. Phys. **3** 566-578

Pais A 1962 *On spinors in n dimensions* J. Math. Phys. **3** 1135-1139

Pauli W 1927 *Zur Quantenmechanik des magnetischen Elektrons* Z. Phys. **43** 601-623

— 1933 *Prinzipien der Quantentheorie* in Handbuch der Physik ed Geiger and Scheel, vol. **24**, pp. 1-278 (Berlin : Springer)

— 1957 *On the conservation of the lepton charge* Nuovo Cimento **6** 204-214

Penrose R 1960 *A spinorial approach to general relativity* Ann. of Phys. (NY) **10** 171-201

— 1967 *Twistor algebra* J. Math. Phys. **8** 345-366

— 1983a *Physical space-time and non-realizable CR-structures* Bull. Amer. Math. Soc. (NS) **8** 427-448

— 1983b *Spinors and torsion in general relativity* Found. of Physics **13** 325-339

— 1987 *On the origins of twistor theory* in Gravitation and Geometry ed W Rindler and A Trautman (Napoli : Bibliopolis)

— and Mac Callum M A H 1972 *Twistor theory : an approach to the quantization of fields and space-time* Phys. Rep. **6 C** 241-316

— and Rindler W 1984 and 1986 *Spinors and space-time* vols 1 and 2 (Cambridge : C.U.P.)

Popov V L 1977 *A classification of spinors of dimension fourteen* Usp. Mat. Nauk **32** 199-200

Porteous I R 1981 *Topological geometry* 2nd ed. (Cambridge : C.U.P.)

Regge T 1984 *The group manifold approach to unified gravity* in Relativity, groups and topology II ed B S Dewitt and R Stora (Amsterdam : North-Holland)

Robinson I 1961 *Null electromagnetic fields* J. Math. Phys. **2** 290-29

— and Trautman A 1986 *Cauchy-Riemann structures in optical geometry* in Proc. 4th Marcel Grossmann Meeting ed R Ruffini (Amsterdam : Elsevier)

Rodrigues O 1840 *Des lois géométriques qui régissent les déplacements d'un système solide* Journal de Math. (Liouville) **5** 380

Salam A and Strathdee J 1982 *On Kaluza-Klein theory* Ann. of Phys. (NY) **141** 316-352

Serre J P 1955 *Applications algébriques de la cohomologie des groupes. Théorie des groupes simples* Sém. H. Cartan 3 année 1950/51 (Paris : Ecole Normale Supérieure)

Shaw W T 1985 *Twistors, minimal surfaces and strings* Class. Quant. Grav. **2** L113-119

Steenrod N 1959 *The topology of fibre bundles* (Princeton : P.U.P.)

Study E 1903 *Geometrie der Dynamen* (Leipzig : Teubner)

Thirring W 1972 *Five-dimensional theories and CP. violation* Acta Phys. Austr. Suppl. **9** 256-271

Tits J 1959 *Sur la trialité et certains groupes qui s'en deduisent* IHES Publ. Math. **2** 13-60

Trautman A 1984 *Differential geometry for physicists* (Napoli : Bibliopolis)

— 1985 *Optical structures in relativistic theories* Astérisque hors série 401-420

— 1987 *Dirac and Chevalley spinors: a comparison* The Trieste "Running Seminar on Spinors" Letter n° 5 (8 October 1987)

Uhlenbeck G E and Goudsmit S 1925 *Spinning electrons and the structure of spectra* Nature **117** 264

Van der Waerden B L 1929 *Spinoranalyse* Nachr. Ges. Wiss. Göttingen math.-phys. **1929** 100

— 1960 *Exclusion principle and spin* in Theoretical Physics in the Twentieth Century : A Memorial Volume to Wolfgang Pauli ed M Fierz and V F Weisskopf (New York :

Interscience)

Wall C T C 1964 *Graded Brauer groups* J. reine und angew. Math. **213** 187-199

Ward R S 1977 *On self-dual gauge fields* Phys. Lett. **61 A** 81-82

Weinberg S 1983 *Charges from extra dimensions* Phys. Lett. **125 B** 265-269

Weiss E 1933 *Oktaven, Engelscher Komplex, Trialitätsprinzip* Math. Z. **44** 580-611

Weyl H 1929 *Elektron und Gravitation. I* Z. Phys. **56** 330-352

— 1939 *The classical groups* (Princeton : P.U.P.)

Witten E 1981a *A new proof of the positive energy theorem* Comm. Math. Phys. **80** 381-402

— 1981b *Search for a realistic Kaluza-Klein theory* Nucl. Phys. **B 186** 412-428

Yang C N 1987 *Square root of minus one, complex phases and Erwin Schrödinger* in Schrödinger ed C W Kilmister (Cambridge : C.U.P.)

GPSR Compliance

The European Union's (EU) General Product Safety Regulation (GPSR) is a set of rules that requires consumer products to be safe and our obligations to ensure this.

If you have any concerns about our products, you can contact us on ProductSafety@springernature.com

In case Publisher is established outside the EU, the EU authorized representative is:

Springer Nature Customer Service Center GmbH
Europaplatz 3
69115 Heidelberg, Germany

Batch number: 10704763

Printed by Printforce, the Netherlands